T0306110

Data Science for the Geosciences

Data Science for the Geosciences provides students and instructors with the statistical and machine-learning foundations to address Earth science questions using real-world case studies in natural hazards, climate change, environmental contamination, and Earth resources. It focuses on techniques that address common characteristics of geoscientific data, including extremes, multivariate, compositional, geospatial, and space–time methods. Step-by-step instructions are provided, enabling readers to easily follow the protocols for each method, solve their geoscientific problems, and make interpretations. With an emphasis on intuitive reasoning throughout, students are encouraged to develop their understanding without the need for complex mathematics, making this the perfect text for those with limited mathematical or coding experience.

Students can test their skills with homework exercises that focus on data scientific analysis, modeling, and prediction problems, and through the use of supplemental Python notebooks that can be applied to real datasets worldwide.

Dr. Lijing Wang is a Postdoctoral Research Fellow in the Earth and Environmental Sciences Area at Lawrence Berkeley National Laboratory. She earned her Ph.D. from the Department of Earth and Planetary Sciences at Stanford University. Her research centers on integrating geoscientific data, such as geophysical surveys and in-situ hydrological measurements, with hydrological modeling to develop informed solutions for water resource management. She was a Stanford Data Science Scholar and had been a teaching assistant for the Data Science for Geosciences course at Stanford for over three years. She has received the Harriet Benson Fellowship from Stanford for her exceptional research accomplishments.

Dr. Zhen Yin is a research scientist and co-founder of the Stanford Mineral-X (https://mineralx. stanford.edu/) at Stanford Doerr School of Sustainability. His research developments in data science have been implemented in various subjects, including Antarctica topographic modeling, critical mineral explorations in Asia/North America/Africa, and North Sea projects. He was previously a Research Associate at the Edinburgh Time-Lapse Project in Scotland, leading a research collaboration with Equinor from 2016 to 2018, and a Postdoctoral Fellow at the Stanford Chevron Center of Research Excellence from 2018 to 2021.

Dr. Jef Caers is Professor of Earth and Planetary Sciences in the Stanford Doerr School of Sustainability. He is the author of a wide range of journal papers across mathematics, statistics, geological sciences, geophysics, engineering and computer science, and four other books. He has received several best paper awards, as well as the 2001 Vistelius Award from the International Association for Mathematical Geosciences (IAMG) and the 2014 Krumbein Medal from the IAMG for his career achievement.

Data Science for the Geosciences

Lijing Wang

Lawrence Berkeley National Lab, California

Zhen Yin

Stanford University, California

Jef Caers

Stanford University, California

 CAMBRIDGE
UNIVERSITY PRESS

Shaftesbury Road, Cambridge CB2 8EA, United Kingdom

One Liberty Plaza, 20th Floor, New York, NY 10006, USA

477 Williamstown Road, Port Melbourne, VIC 3207, Australia

314–321, 3rd Floor, Plot 3, Splendor Forum, Jasola District Centre, New Delhi – 110025, India

103 Penang Road, #05–06/07, Visioncrest Commercial, Singapore 238467

Cambridge University Press is part of Cambridge University Press & Assessment, a department of the University of Cambridge.

We share the University's mission to contribute to society through the pursuit of education, learning and research at the highest international levels of excellence.

www.cambridge.org
Information on this title: www.cambridge.org/highereducation/isbn/9781009201414

DOI: 10.1017/9781009201391

First published 2024

A catalogue record for this publication is available from the British Library.

Library of Congress Cataloging-in-Publication Data
Names: Wang, Lijing, 1995– author. | Yin, Zhen, 1989– author. | Caers, Jef, author.
Title: Data science for the geosciences / Lijing Wang (Stanford University, California), Zhen Yin (Stanford University, California), Jef Caers (Stanford University, California).
Description: Cambridge, United Kingdom ; New York, NY : Cambridge University Press, 2024. | Includes bibliographical references and index.
Identifiers: LCCN 2023000678 | ISBN 9781009201414 (hardback) | ISBN 9781009201407 (paperback) | ISBN 9781009201391 (ebook)
Subjects: LCSH: Earth sciences – Statistical methods. | Earth sciences – Data processing.
Classification: LCC QE33.2.S82 W36 2024 | DDC 550.285–dc23/eng20230512
LC record available at https://lccn.loc.gov/2023000678

ISBN 978-1-009-20141-4 Hardback
ISBN 978-1-009-20140-7 Paperback

Additional resources for this publication at www.cambridge.org/ds4gs

Contents

The plate section can be found between pp 114 and 115

Preface

Motivation

Since the start of this century, data collection to study our Earth has seen an incredible expansion. At the same time, the survival of our species as well as the planet-wide ecosystems has emerged as a critical focus: climate change, sustainability, and energy transition are now part of our daily language. These three urgent challenges facing humanity cannot be solved without geosciences. Geoscience programs have grown, their methodologies increasingly using data science, machine learning, and artificial intelligence as critical tools to study, understand, and eventually aid in mitigating the imminent threats we are facing.

What has received less urgent attention is the development of educational programs around data science with a focus on the geosciences. This poses a problem. Many students still take their first data science class in a statistics, engineering, or computer science department, where the geosciences may not get much attention. Economy, engineering, medicine, finance, or business applications are often the focus. Yet, the challenges in analyzing and predicting with data in the geosciences are often more difficult because of the very nature of the data. Our data are extremal, multivariate, chemical, physical, spatial, temporal, and multiscale as well as usually being massive and highly uncertain and biased. Careful handling of basic physical principles is required, so that mass balance is not violated and the stoichiometry of the geochemistry involved is accounted for. This textbook focuses on data science methodologies of analysis and prediction that are specifically relevant for the geosciences, and often not covered in traditional courses, which aim for foundational and broad coverage rather than geoscientific applications. Each chapter starts from real problems with real datasets provided to us by geoscientists, engineers, and decision makers working on them. Therefore each one also has a human focus: humans in their relationships to hazards, water, food and energy resources, environmental contamination, and climate change. We believe this will make the material highly relevant and highly attractive to the current generation of students who want to make an impact. In order to do this, students should become familiar with society-relevant problems, as well as specific methodologies and practical data-scientific tools to address those challenges. Our focus therefore is to start with a problem, then use these tools to study and contribute to it.

Structure of the Book

To achieve this stated goal, we have compiled four key areas of data science from a geoscience perspective: (1) extreme value statistics, (2) multivariate analysis, (3) spatial data aggregation, and (4) geostatistics.

Each chapter has the following key pedagogical components:

- desired learning outcomes at the start of each chapter
- within-chapter check-ins, summarizing what should have been learned
- protocols for executing specific tools to aid students in structuring data analysis
- end-of-chapter "test your knowledge" questions
- annotated further reading lists

We now explain the motivation behind the choice of these four topics, as well as the philosophy in covering them.

Chapter 1 focuses on predicting extremes larger than observed in datasets. An important area of application is natural hazards. In this chapter, we use volcano eruptions as a specific example. We start by focusing on graphical techniques, in particular quantile–quantile plots to analyze extremal data. We show why the exponential quantile plot is an essential tool in extremal analysis. We also focus on the generating mechanisms and underlying process for extreme magnitudes: (1) proportional growth leads to lognormal models and (2) fractal growth leads to Pareto distributions. We then pose the challenge of picking a suitable probability distribution model to estimate extremes. This is the topic of extreme value theory. We do not make derivations of theoretical models; instead, we illustrate how these models emerge from intuitive Monte Carlo experiments. We also emphasize that a key statistical parameter emerges: the extreme value index. We then link this index to quantile plots such as the Pareto quantile plot and the mean excess quantile plot. We conclude with practical examples of predicting rare large diamonds, as well the return period of large volcano eruptions from a dataset that is limited in time. Extreme value analysis is likely a more challenging topic of the book because of its significant mathematical background. We mitigate this potential difficulty by continuously linking theoretical concepts to quantile–quantile plots. We focus on interpreting such plots and showing how they can be used to validate any numerical prediction.

Chapter 2 focuses on the most essential tools for analyzing compositional and/or multivariate datasets that often emerge when performing geochemical analysis. It starts by introducing groundwater contamination in one of the world's largest agricultural areas: the Central Valley of California. Using this case, we introduce the purpose of data science: discovering the processes that caused contaminations, whether geogenic or anthropogenic. Knowing the causes helps to mitigate against them. We take you on a path of discovery through several protocols of applying data-scientific tools to unmask the processes. The key to using these tools is to understand the compositional nature of geochemical datasets, and how compositions need to be treated appropriately to draw meaningful conclusions, a field termed compositional data analysis. Two dual methods of dimensionality reduction are covered, the well-known principal component analysis and the lesser-known multidimensional scaling. Our approach to developing these techniques is guided by intuition using geometric analysis. Significant contamination can be detected using multivariate outlier detection, an important area of data science that looks at unmasking anomalies in complex multivariate datasets. After extensive data analysis, we introduce factor analysis, a powerful, but simple statistical technique for predicting causes.

This chapter emphasizes the need for data scientists to work with domain experts. Geochemical datasets are complex because of the intricate chemical reactions that have taken place. Hence the data are the outcome of chemical processes, and they are used to discover these processes. We emphasize in this chapter, and throughout the book, that any good data science will require collaboration with domain experts. Our teaching emphasizes this by inviting domain experts to the course, and have students ask informed questions about what in the datasets they should focus on, while the domain expert plays the role of a non-expert data scientist. We recommend this approach as an educational component that highlights how data science collaborations work in postgraduate careers.

Chapter 3 integrates heterogeneous spatial or spatio-temporal data sources, with the aim of predicting the occurrence of hazards and resources. Much of the transition to the Net Zero Emissions by 2050 Scenario will rely on changing from fossil fuels to materials for renewable energy and batteries. Mineral exploration is therefore key to achieve this net zero carbon dioxide emissions goal. Readers will engage in an active mineral exploration for battery metals in Cape Smith, Canada, executed by an actual mineral exploration company in 2021–2022, and will learn how data science can be an effective guide in geological fieldwork. To achieve this, several spatial information sources are used, such as remote sensing, geophysical information, and geochemical information. These sources need to be aggregated to guide field geologists in locating areas of interest with the aim of collecting samples. In that context, we introduce Bayes' rule and Bayesian reasoning about knowledge and information. We show how essential Bayes' rule is in reducing uncertainty about occurrence with heterogeneous data sources. We emphasize the counterintuitive results of Bayesian reasoning: rare events are very difficult to predict even with very accurate data. Here we make the link with exploring resources, which entails locating rare events or anomalies. Next, we cover the alternative to Bayes' rule: logistic regression. Regression is used to model the uncertainty of some rare events from information sources. We emphasize the advantage and disadvantage of these opposite approaches. Bayes' reasoning and logistic regression are also foundational methods in machine learning. Coverage of further machine learning methods is beyond the scope of this short textbook. Instead, we emphasize that many machine learning methods find their origins in statistical methods.

Chapter 4 provides tools for spatio-temporal data analysis and predicting in space–time. The origin of geostatistics lies in subsurface applications, starting with mineral deposits and mining. The subsurface application we cover in this book is sustainable farming in Denmark. We introduce geophysical techniques for imaging the geochemistry of the subsurface. The challenge here lies in combining information from wells with geophysical data. Readers will learn this is a common problem in modeling the subsurface: sparse accurate data with exhaustive indirect data. A second application happens at the surface of the Earth: glaciers melting in Antarctica. We discuss the significant ongoing effort in mapping, using radar imaging, the Thwaites glacier in Antarctica–a glacier that is key to predicting sea-level changes over the next 100 years. Both cases, Denmark and Antarctica, call for building spatial models from incomplete data: spatial interpolation. We cover geostatistical methods for capturing spatial variability with variograms. Readers will learn this in an intuitive way by means of images: what they can visually observe in image datasets and how this relates to mathematical quantifications, such as variance, correlation length, or trends. We cover spatial interpolation, also known as kriging, without full mathematical derivations. Instead, we propose an intuitive way of showing why variograms are essential to spatial interpolation. We illustrate how variograms are measures of distance between locations in space, and how such distances are essential to creating interpolation that does not show any artifacts. Next, we show the limitation of kriging, namely, that it creates interpolations that have less variance than the dataset used for interpolation. We emphasize with practical examples why that is important, which allows us to introduce the concept of conditional simulation where the aim is to generate many interpolated maps that reproduce real variation. We show how these maps represent spatial uncertainty, and how that uncertainty affects prediction, such as predicting groundwater redox conditions in the subsurface of Danish agricultural areas. Finally, we introduce ways of interpolating by using training images, which are datasets that show what the expected spatial variability is. We highlight how this methodology is relevant in Antarctica: the exposed Arctic topography can help us interpolate Antarctica.

As the book is aimed at any student interested in data science applications in geosciences without requiring any additional background in mathematics or statistics, we have added a final Chapter 5, which introduces the topic of concept reviews. This includes some foundational materials such as probability densities, Monte Carlo methods, and Bayes' rule. Our approach is to emphasize why a particular technique is used, and how to interpret plots and tables generated by these techniques within the context of a relevant problem. For the same reason, we aim to generate first an intuitive understanding of the technique; then, if the student is interested, they can dive deeper into the methodology. Starting from a strong mathematics focus would also make it challenging to cover a lot of ground. For example, principal component analysis can be taught by deriving the equations and making the link with eigenvalue decomposition of the covariance matrix. Instead, we start from simple two- and three-dimensional datasets and appeal to the student's insight into the geometric aspect: the study of an ellipse, and how we can transform it to a circle. This geometric aspect is explained without equations, but instead with plots and figures that appeal to intuition, starting from geometry. In general, it is our experience that students in the geosciences retain much more practical knowledge when presented with material from case studies and intuitive reasoning compared to using rigorous mathematical proofs or existing coding techniques.

The Audience

The material presented in this textbook is taught at Stanford in two courses: an advanced level course for undergraduates and a course for those starting graduate school. Hence, for undergraduates we cover less material but with more fundamental concepts. For graduates, we cover material faster, and at a more advanced level. Regardless, any instructor interested in using this resource may choose for themselves what material they feel should be emphasized or can use it to possibly develop their own course. We feel that one of the significant elements of this resource lies in the societal-relevant datasets. The cases and datasets are significant and substantial. They cover many aspects of the geosciences, from mineralogy to environmental science; from volcano eruptions to melting glaciers. As such, the material also provides students and instructors with an introduction to a geoscience field they might be less familiar with.

Supplementary Material

Opensource collaboratory notebooks are available for students to use, containing algorithms presented in the textbook material (see www.cambridge.org/ds4gs). These notebooks also provide starting code for students doing homework assignments. We feel that this way the student does not have to worry about being able to write their own code. Since the aim of our book is to teach practice, interpretation, intuition, and societal relevance, coding would take away valuable time spent on these learning elements. These notebooks also provide equal opportunities to those who are not strong in coding.

Additionally, we have created lecture videos on key concepts covered in each chapter, including the usage of Stanford Geostatistical Modeling Software (SGEMS) in Chapter 4, and we have created a short Compositional Data Package (CoDaPack) tutorial on how to use this software for compositional data analysis in Chapter 2. The intended learning outcome is for students to be able to explain intuitively and concisely to a non-expert what important data scientific concepts are.

For instructors, we share the lecture slides based on this material, homework assignments that are linked to the notebooks, and answers to the "Test your knowledge" question.

Acknowledgements

It was with great pleasure that we worked on this book, and we have many to thank for their contributions. Many at Kobold Metals – Kurt House, I-Kang Ding, Mark Topinka, and Peter Lightfoot – were instrumental to the mineral exploration case. We thank Alandra Lopez for providing the Central Valley water quality dataset and the many interesting discussions around interpreting factor analysis. We thank Chen Zuo, Chang'an University, and Emma "Mickey" MacKie, University of Florida, for collaborating on modeling the Thwaites and Pine Island glaciers. We thank Hyojin Kim and Birgitte Hansen, Geological Survey of Denmark and Greenland, as well as Troels N. Vilhelmsen and Anders V. Christiansen at Aarhus University for their willingness to share the geochemical and tTEM data as well as their insights into the groundwater challenges in Denmark. We would also like to thank the teaching assistants, Tyler Hall and Yizheng Wang, and the students of GEOLSCI 6 and GEOLSCI 240 for their interest and feedback on the course materials. We thank Sarah Strange of Cambridge University Press for her expert guidance in creating audience-relevant contents. We'd like to thank the many students, faculty, and collaborators involved in the Stanford Center for Earth Resources Forecasting and Stanford Mineral-X, and for their continued support in making our planet a better place to live in.

1   Extreme Value Statistics

Expected Learning Outcomes

- You will learn that problems involving extreme values require a specialized treatment.
- You will learn about size distributions in nature and how they are linked to physical/chemical processes.
- You will learn ways to visualize extremes in a dataset and perform exploratory analysis using quantile plots.
- You will learn to make predictions of extreme values even larger than those occurring in your data, both in terms of magnitude and frequency.

1.1   Introduction

In the geosciences, we often study occurrences of extreme events. Extreme highs (or lows) can be of scientific interest, or directly affect our daily lives, such as in the case of floods or earthquakes. Extreme value statistics is a branch of applied statistics that provides ways of analyzing and predicting such high values. Why do we need a new form of statistics, and just not "regular" statistics? For one, we are interested in extrapolating beyond the highest value in a dataset. Because sampling is limited (either in time or space), it is unlikely that we have observed all high values. We'd like to address questions such as: if we have accurately recorded the magnitude of floods in some area over the last 50 years, what magnitude of floods can we expect over the next 100 years? There is a high probability that there will be a flood in the next 100 years that is larger than the maximum one observed over the last 50 years. Extrapolation should be done with care and have a clear foundation. Extrapolating in statistics and machine learning is fraught with difficulties: there is no guarantee that the statistical models doing interpolations apply equally for unobserved large values. Extreme value theory provides rigorous ways of reasoning about extrapolation, and this branch of statistics employs systematic procedures for predicting extreme values, based on that theory.

Before heading into the methodologies, a note of caution: high values can be named as many things, such as outliers, anomalies, or extremes. Technically, at least according to principles of statistical sciences, these terms define very different things. Outliers and anomalies are different

Figure 1.1 The difference between an extreme and an outlier/anomaly.

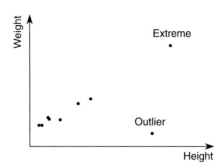

from extremes. In short, an extreme is something that can be predicted by extrapolating from the population of lower values in a sample dataset. In statistical terms, it is part of the population of all values: low, middle, high, and extreme. They all form one happy family. An anomaly or outlier is something that behaves completely differently from the "background" population. Consider a simple example in Figure 1.1. The "extreme" in the figure can be treated as an extrapolation from the main height and weight population. But the "outlier" cannot. Clearly, distinguishing between the two in real datasets is very challenging. There will rarely be full certainty whether the highest sample value is an extreme or an anomaly, but we can put some measure of confidence on it. In this chapter, we provide some visual tools for determining outliers versus extremes in the univariate case (just one variable). In Chapter 2 we extend that to the multivariate case (two or more variables, as in Figure 1.1).

1.2 Motivating Examples

The geoscientific examples most applicable in the context of extremes are related to "size." Size here can mean many things: the size of an object (craters, diamonds) or the magnitude of an event (floods, volcanic eruptions). Size distributions are well studied in the physical and statistical sciences. In the geosciences context, size distributions are the result of the physical/chemical processes and the conditions under which they take place. We cover two examples of size in the general sense: the size of diamonds and the magnitude of volcanic eruptions.

1.2.1 Diamonds

Diamonds are formed in the deep subsurface under ideal pressure and temperature conditions, then brought to the surface through kimberlite (a type of ancient volcano) eruptions. The speed of these eruptions is important as diamonds are not at thermodynamic equilibrium at surface conditions, so they are sort of "frozen" in their state. This is why you should not heat up diamond jewelry to about 763 °C, they will become carbon dioxide! The size distributions of diamonds in a kimberlite are therefore a function of the temperature, pressure, and carbon concentration when formed. But since we do not know these conditions today, we only observe the resulting size distribution. Exploration for diamonds therefore consists first in finding kimberlite pipes although not all kimberlites have commercial-sized diamonds. Another source for diamonds is when kimberlites are eroded and the materials transported along rivers (and even beaches, as in Namibia). The roughness of the river bedrock causes the diamond to be trapped, forming what are called placer deposits.

Once diamond deposits are found, one needs to determine whether they are economically viable for mining, hence a valuation is required. This requires sampling, which means drilling or excavating part of the deposits and analyzing the diamonds recovered. Because diamonds are rare, only a few hundred to a few thousand stones are recovered at the initial sampling stages. These stones are weighed, and their dollar values are calculated from attributes such as quality, inclusion, color, etc. . . . Diamond weight is expressed in carats (ct). Hence, for every stone we have two variables: size and dollar value. We will consider here only the size. Since large stones are very valuable (e.g., 2 ct stones fetch $5000 to $50 000 ;10 ct between $20 000 and $200 000 in 2020), one would like to know how many large stones are present in the entire deposit, given a limited sample. For example, if we mined 100 000 stones, how many stones are larger than 2 ct? Clearly, we need to know the probability of the size being larger than 2 ct, and hence we need to know the size distribution (see Section 5.2.2). A rather simple way would be to do this empirically, for example just counting the number of stones in the sample above 2ct, then estimating a probability by dividing that count by the total number of stones. Two problems exist with this approach. One is very practical, and the other is statistical. Since exploration companies are not interested in small stones, say less than 0.1 ct, they tend to use sieves that disregard those stones; therefore, it is very likely that we would get progressively less of the smaller stones than if we did a full analysis. Hence, the total number of small stones is not really a trustworthy number. Secondly, an empirical estimate is subject to sampling uncertainty. The smaller the probability we are trying to assess, the more difficult it becomes to assess it. Indeed, consider a hypothetical dataset of diamonds that has 3 out of 1000 stones larger than 2 ct. Consider, by chance (not a very unimaginable one!), that we did not observe the largest stone (just missed it), then the probability of stones being larger than 2 ct (the probability of exceedance) goes from 3/1000 to 2/1000. This would mean that your revenue estimate will go down by at least 33%! The uncertainty is substantial. The empirical approach will stop working altogether when we choose to count anything larger than the largest sample value.

What are the practical questions? A mining company would like to know how many diamonds will be mined above a large threshold, such as 10 ct, relative to the amount of volume of rock/soil that will be excavated.

 Visit **Notebook 01: Diamonds** to download two diamond datasets and perform some exploratory data analysis.

1.2.2 Magnitudes of Natural Events

As mentioned before, sizes can be in the form of magnitudes, with examples such as volcanic eruptions, earthquakes, rainfall, and landslides. These geological events are quite complex in terms of their physical attributes. To summarize such events, geoscientists have defined

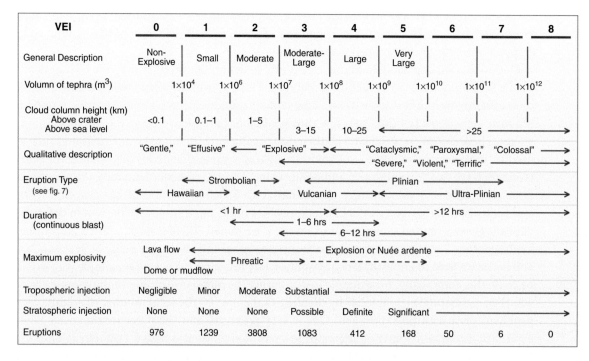

Figure 1.2 Description of Volcanic Explosivity Index (VEI) (reprinted with permission from Newhall and Self, 1982; Siebert et al., 2010)

common scales in order to compare them across many regions of the world. Here we focus on large volcanic eruptions.

Databases of volcanic eruptions typically list values such as tephra, eruption age, intensity, magnitude, column heights, and volcanic explosivity index (VEI) (see Figure 1.2). Magnitude and intensity are defined as follows:

$$\text{Magnitude} = \log_{10}[\text{erupted mass, kg}] -7$$

$$\text{Intensity} = \log_{10}[\text{mass eruption rate, kg/s}] + 3$$

Compiling these databases (Figure 1.3) is a challenge since accurate actual measurements have been made only recently. For magnitudes going back hundreds or thousand of years, only indirect data is available, such as from ice cores. As with diamonds, large events are more likely recorded than small events.

What are practical questions here? For one, we would like to know the relationship between magnitude and frequency (in time). Unlike diamonds, the time component now becomes important. How often do volcanic eruptions with magnitude larger than 7 occur? They do not occur exactly with the same time interval, so what is the variation of the length of time between any two magnitudes larger than 7? There are two ways to deal with time: work on the exact occurrence in time (some distinct event/point in time); or work with time intervals, which is a continuous variable. Then, we can turn to questions such as: What is the probability of having a magnitude large than 7 (over all time)? Finally, a very intriguing question: What the maximum possible

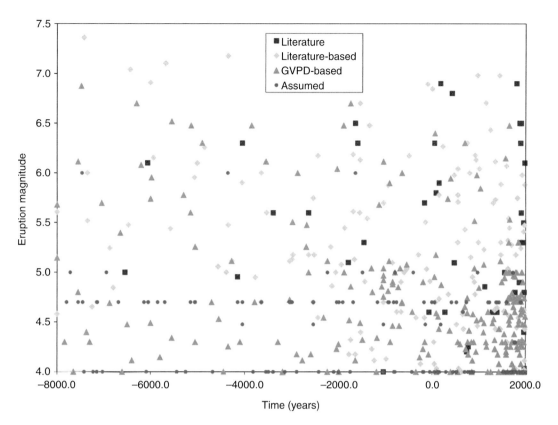

Figure 1.3 Database of major volcanic eruptions over the last 10 000 years (GVPD is the Smithsonian Global Volcanism Program Database). (Reprinted with permission from Deligne et al., 2010)

volcanic eruption? That maximum number must be finite, hence any distribution that goes from zero to infinity is now excluded, meaning normal, lognormal, exponential, etc. What distributions have a finite maximum? Extreme value theory provides models for such cases.

 Visit **Notebook 02: Volcano** to download the volcano magnitude catalogue and perform some exploratory data analysis.

1.3   Concept Review

In case you need a refresher, or are new to data science and would like an introduction to basic concepts, consult Chapter 5, and review the material in Section 5.2 more specifically, to learn about:

- an intuitive way to think about the concept of probability density functions
- the central role the exponential probability density model takes in data science and in particular extreme value analysis

- how Monte Carlo simulation can be used to study the properties of probability distribution models
- hypothesis testing using the bootstrap method: hypothesis testing is key to many areas in science

1.4 Probability Distribution Models for Magnitudes

To make predictions, we need mathematical models. Mathematical models appear everywhere in science. For example, we can describe wave forms as a combination of sine and cosine functions; then we use these functions to predict arrival times of future waves. These functions are mathematical models; they attempt to describe reality as accurately as possible. The same idea applies to datasets of a single variable, such as diamond size. Before doing so, in this section, we will look at various ways to display a dataset, such that it provides us with information as to which mathematical model is most appropriate. These mathematical models are termed probability density functions, covered in this section, as well as more extensively in Section 5.2.2, for those who need a detailed refresher.

1.4.1 The Histogram

The usual way of presenting data is to plot a histogram. In a histogram, we create bins (intervals) and count how many times we find values in each bin. In this way, we turn continuous values into discrete frequencies that we can plot in a single graph. Let's consider that we have two datasets of diamonds, each from a very different deposit. The histograms of both datasets are shown in Figure 1.4. We notice that each dataset has many small stones and a few large stones. The data is termed to be "skewed" (to the right). This is common with many geoscientific datasets because many physical and chemical properties are positive, for example, frequency, resistance, density, proportion, reaction rates, etc.

The histogram is limited in showing details of the dataset. Because of the binning, it is also challenging to compare two datasets just by looking at their histograms. Hence, to get a better understanding, we calculate some summary statistics. Typical statistics that are displayed next

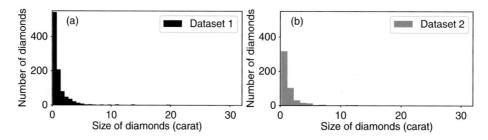

Figure 1.4 Histograms of two diamond datasets, with the size of a diamond measured by its weight in carats: (a) dataset 1; (b) dataset 2.

to the histogram are the maximum, the minimum, the mean, and the standard deviation. Consider that we call our sample set $x_1, x_2, \ldots x_n$, then the arithmetic mean is:

$$\bar{x} = \frac{1}{n} \sum_{i=1}^{n} x_i \qquad (1.1)$$

and the empirical variance

$$s = \frac{1}{n} \sum_{i=1}^{n} (x_i - \bar{x})^2 \qquad (1.2)$$

The mean indicates the center of the data, although in the case of very skewed distributions this center may not be very meaningful. The variance would typically indicate the extent to which data are spread around the mean, but again this would only make much sense if the histogram of the data is somewhat symmetric. In addition, a problem occurs with calculating the mean and standard deviation in the presence of large values. Consider in dataset 2 that we remove the three largest stones. Imagine that, by pure chance, we missed them in our sampling of the deposit. Let's recalculate the mean and standard deviation; see Table 1.1.

We notice significant changes in these statistics when we miss a few high values. In statistics, we would say that these figures are not robust, meaning that they are sensitive to changes in the data, and hence not very reliable.

1.4.2 The Quantile–Quantile Plot

It is clear from the previous section that we need better tools than histograms and summary statistics to look at data when dealing with magnitudes. Here, we introduce the quantile–quantile plot or QQ plot. What is a quantile (see also Section 5.2.1)? To explain this concept, let's consider two simple datasets, with arbitrary units, shown in Table 1.2. Are these datasets similar or different?

To get this insight, we first rank the values in each dataset. In dataset 1, the smallest number is 7. That means only one value is less than or equal to 7, and that is 7. Since we have six values, there is a 1/6 chance that by random drawing (see Section 5.2.3) from this dataset we would find a value less or equal to 7. We term the value 7 the 1/6 quantile of that dataset. It is termed an empirical quantile because we get it from measured data. Later, we will discuss theoretical quantiles, namely those obtained from models.

Table 1.1 Mean and variance of dataset 2

	Mean	Variance
Dataset 2	1.56 ct	5.7
Dataset 2, removing the three largest stones	1.44 ct	3.08

Table 1.2 Two simple datasets, ranked

Dataset 1	34	21	8	7	10	15
Dataset 2	16	22	5	9	11	37
Rank order						
Dataset 1	7	8	10	15	21	34
Dataset 2	5	9	11	16	22	37
Percentile	1/6	2/6	3/6	4/6	5/6	6/6

Figure 1.5 An example of a QQ plot between two simple datasets with six samples. The line drawn is the 45-degree line.

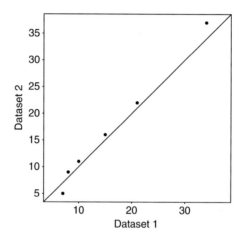

What is the 1/6 quantile in dataset 2? The answer is 5. Let's look at the next quantile, which would be the 2/6 = 1/3 quantile; for dataset 1 this is 8 and for dataset 2 it is 9. How many quantiles can we calculate? As many as we have samples, namely six.

A QQ plot between two datasets is a comparison between the empirical quantiles of each dataset, or it is a plot that compares the ranked values of each dataset. This is shown in Figure 1.5. If these points fall near a 45-degree line, then we can state that the datasets are similar in distribution. In QQ plots we look for alignment, not necessarily exactly along the 45-degree line. Take for example dataset 2 and add some constant to all values; in other words, we increase the mean. The result is that the dots in Figure 1.5 will shift up. Now consider multiplying all values with a constant. If this constant is larger than 1, we increase the variance of the data. The result of the multiplication is that the dots in the QQ plot will rotate away for the 45-degree line, but still form a line. In QQ plots, we can diagnose similarity in distribution even if datasets have different means and variances.

An alternative way to make the same QQ plot is by using empirical cumulative distribution functions (see also Section 5.2.2). That's a long description for making a staircase-like function with the data, shown in Figure 1.6a, as follows:

- rank a dataset with n samples from small to large
- plot the data by making a step of size $1/n$ on the y-axis

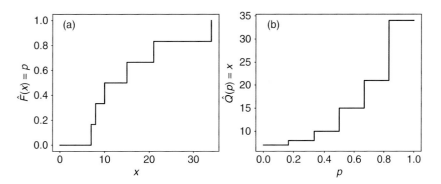

Figure 1.6 (a) Empirical cumulative distribution function (staircase function) and (b) empirical quantile function (inverse staircase function) for dataset 1 in Table 1.2.

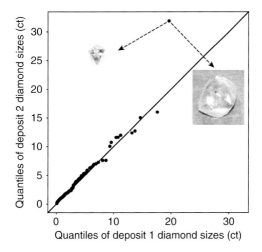

Figure 1.7 QQ plot between two parcels of diamonds. The insets show photographs of one big diamond from deposit 2 and a small diamod from deposit 1. (Smaller diamond, photographed by Rob Lavinsky, distributed under a CC BY-SA 3.0 license; larger diamond, photographed by Eurico Zimbres FGEL/UERJ, distributed under a CC BY-SA 2.0 br license.)

Since we accumulate steps of size $1/n$, this is a cumulative representation of the data, hence the term empirical cumulative distribution function (empirical = measured data). On the y-axis we have percentages p; on the x-axis, the data values. Using any p-value, we can find (by going in reverse = the inverse) the corresponding x-value. Also, we can switch the axis and make the y-axis the x-axis, giving an inverse staircase function, which is termed the empirical quantile function. In data science, we denote empirical cumulative distribution function (staircase function) as $\hat{F}(x) = p$, where the hat means "empirical" (without the hat it means "model" or "theoretical"; see also Section 5.2.2). Likewise, the empirical quantile function is written as $\hat{Q}(p) = x$.

Figure 1.7 shows the QQ plot of the two diamond datasets. We observe that the quantiles between the two datasets are comparable, except for the largest values. The question now is whether these two datasets are really different, or if this largest value difference is just by chance. Clearly, we need more tools that allow us to zoom in to what happens in the tail of the distribution, meaning the variation of the larger values.

Note: A QQ plot is not a scatter plot. In a scatter plot we aim to observe whether two datasets are correlated, while in a QQ plot we compare their histograms. Correlation in scatter plot is discussed in the next chapter.

> ▶   Play **Video 01: Quantile Plots and Extreme Value Index** to learn how to make and interpret quantile plots.

1.4.3   The Logarithm

The logarithm is often introduced in calculus as the "opposite" or "inverse" of exponentiation. Exponentiation makes small values larger, while the logarithm makes large values smaller, and hence more comparable to other small values. But there is more to the logarithm than just a calculation trick: it has physical meaning as well. The concept of the logarithm goes a long time back and exists in various cultures, Western and Eastern. Archimedes called the logarithm the "order of a number," 10 has order 1, 100 has order 2. The Indian mathematician Virasena worked with the concept of "Ardhaccheda": the number of times a number could be halved. For example, 2^x could be halved x times (the base 2 logarithm). What was common in these studies is that a logarithm turns a product into a sum, hence allowing for easy calculation. But there is more than meets the eye here. It was found that some human and natural processes work geometrically and not arithmetically. For example, take the investment of money. This process is not additive but multiplicative. It means that the profit of such money can be reinserted into new investments, hence compounding, or creating a multiplicative effect. This is a good thing, otherwise, retirement portfolios would grow very slowly! As a consequence, we need to create an alternative to the arithmetic mean and consider the average multiplicative effect:

$$\overline{g} = \sqrt[n]{\prod_{i=1}^{n} x_i} \tag{1.3}$$

where $\overline{g}$ is termed as the geometric mean; $\prod_{i=1}^{n} x_i$ means multiplication from x_i to x_n. Note that the geometric mean can be calculated by taking the log of data and then "exp" (take the exponent) the total:

$$\log(\overline{g}) = \frac{1}{n}\sum_{i=1}^{n}\log x_i \Rightarrow \overline{g} = \exp\left(\frac{1}{n}\sum_{i=1}^{n}\log x_i\right) \tag{1.4}$$

After taking the logarithm, we can recalculate the histograms (see Figure 1.8). We notice how the histograms now become more symmetric, and hence the arithmetic mean of this logarithm (the geometric mean) now makes much more sense.

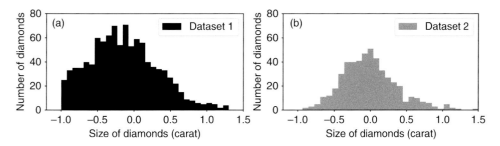

Figure 1.8 Histogram of the logarithm of diamond sizes: (a) dataset 1, (b) dataset 2.

There is an intuitive and physical reasoning that explains why the logarithm is often associated with "size." If you think about size as something that is created because of "growing" a process, such a process of change or increase can be seen as a multiplicative effect. The larger something is, the more chance there is to add more "size," hence we can hypothesize the following proportional effect:

$$X(t) - X(t-1) = \varepsilon(t)X(t-1) \tag{1.5}$$

where t represent the growing process (in time) and $\varepsilon(t)$ represents the proportional effect. If we develop this further and add all proportional effects together:

$$\frac{X(t) - X(t-1)}{X(t-1)} = \varepsilon(t) \;\Rightarrow\; \sum_{t=1}^{n} \frac{X(t) - X(t-1)}{X(t-1)} = \sum_{t=1}^{n} \varepsilon(t) \tag{1.6}$$

Here is where the logarithm appears:

$$\sum_{t=1}^{n} \frac{X(t) - X(t-1)}{X(t-1)} \simeq \int_{X(0)}^{X(n)} \frac{\mathrm{d}X}{X} = \log\left(X(n)\right) - \log\left(X(0)\right) \tag{1.7}$$

The summation becomes the approximation of integration: the logarithm is the integral of $1/X$. Then Eq. (1.7) shows how the logarithm is the summation of very small proportional effects.

1.4.4 The Bell Shape

After taking the logarithm of data, we notice that the data become symmetric like a bell, a commonly observed shape. It was first observed by the German mathematician Carl Friedrich Gauss, who created a mathematical model to represent this shape, termed the Gaussian distribution model:

$$f_X(x;\mu,\sigma) = \frac{1}{\sigma\sqrt{2\pi}} \exp\left(-\frac{(x-\mu)^2}{2\sigma^2}\right) \tag{1.8}$$

Figure 1.9 Examples of the Gaussian distribution model when varying the mean μ and variance σ^2.

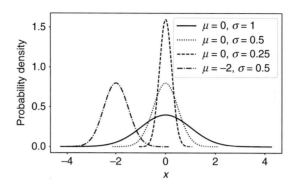

Figure 1.10 Examples of the lognormal distribution model when varying the mean μ and variance σ^2.

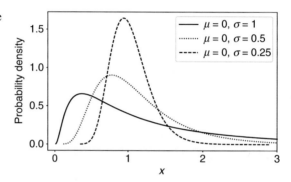

It is also commonly known as the normal distribution model. Equation (1.8) is a probability density function, which is reviewed extensively in Section 5.2.2, in case this concept is new to you. This model has two free parameters, the mean μ (the middle) and the standard deviation σ or variance σ^2 (the square of the standard deviation). In Figure 1.9, we explore a few settings of these new parameters.

We observed this bell shape only after taking the logarithm, so logically, when we revert this operation, we need to take the exponent. As a consequence, this will transform the bell shape into a new shape. This is shown in Figure 1.10.

We notice that all values are positive (indeed, we take the "exp" function). One can also derive a closed-form expression of this curve, namely

$$f_X(x;\mu,\sigma) = \frac{1}{x\sigma\sqrt{2\pi}}\exp-\left(\frac{(\log(x)-\mu^2)}{2\sigma^2}\right) \tag{1.9}$$

This distribution model is termed the lognormal distribution model. It is a very common distribution model in the geosciences, but not the only one, as we will discuss in the next section.

We would often like to know if the data follow a lognormal distribution or not. In the next sections, we discuss quantitative methods to figure this out. Here, we focus on a graphical representation, the quantile–quantile plot.

In previous cases, we compared empirical quantiles of two datasets. Now we turn to comparing the quantiles of a dataset with the quantiles of a theoretical model such as the

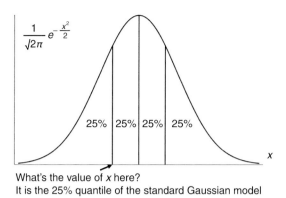

Figure 1.11 Quantiles (25%, 50%, 75% quantiles) of the standard Gaussian distribution.

lognormal distribution. The aim is to investigate visually whether the data follow this model.

What is a theoretical quantile (see also Section 5.2.2)? To understand this, we need to go back to the definition of a quantile. A quantile is a value that is associated with a probability (or percentile). Take for example the Gaussian distribution with $\mu = 0$ and $\sigma = 1$. The mean here is also the middle of the distribution because the distribution is symmetric. Hence, the mean is also the median or 50% quantile. What is then the 25% quantile? In Figure 1.11, it means there is some value, such that the area under the Gaussian curve (i.e., the integral) is 25% of the entire area under that curve. To calculate this, we need a computer algorithm that solves the following problem:

$$\text{find } x \text{ such that } \int_{-\infty}^{x} \frac{1}{\sigma\sqrt{2\pi}} \exp\left(-\frac{(y-\mu)^2}{2\sigma^2}\right) dy = 0.25 \tag{1.10}$$

That algorithm will return the value -0.675. Because of symmetry, the 75% quantile is $+0.675$. If our dataset has $n = 500$ stones, then we have 500 quantiles. After ranking them from small to large, we can associate a probability of $i/(n+1)$, $i = 1, \ldots, 500$. We need to use $n+1$ and not n because $n/n = 100\%$. If we solve the above problem where the integral equals to 100%, we have an infinite value $x = \infty$, and we can't make plots with infinity as a value! That is why we use $n+1$ as the denominator.

We calculate all theoretical quantiles of the standard normal distribution as shown in Table 1.3, which can be converted into a figure. The plot we get, Figure 1.12a, clearly deviates from the 45-degree line, so we can be certain the data do not have a normal distribution. The story changes when we take the logarithm of the data, resulting in the lognormal quantile plot, Figure 1.12b. Except for the smallest diamonds there appears to be a reasonable agreement between the data and the quantiles of the lognormal distribution.

1.4.5 Self-Similarity

The proportional effect is but one way to create size distributions. A second major family of size variation exists. To understand this second way, let us consider measuring the length of the coastline of Great Britain; see Figure 1.13. The challenge with this problem is that the length

Table 1.3 Standard Gaussian quantiles compared to quantiles of diamond dataset 1

Percentiles	Standard Gaussian quantiles	Quantiles for dataset 1
1/501 = 0.199%	−2.88	0.1
2/501 = 0.399%	−2.65	0.1
3/501 = 0.598%	−2.51	0.1
. . . . . .		
499/501 = 99.60%	+2.65	19.75
500/501 = 99.80%	+2.88	19.96

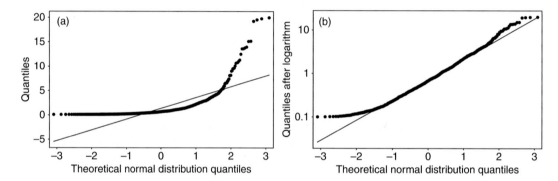

Figure 1.12 (a) Normal quantile plot and (b) the lognormal quantile plot for diamond dataset 1. The straight lines are the 45-degree lines.

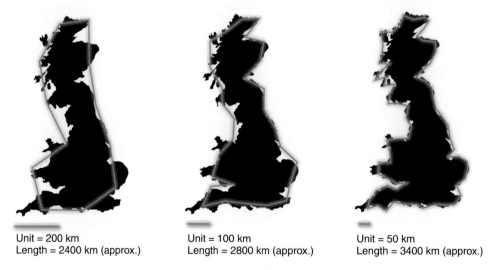

Unit = 200 km
Length = 2400 km (approx.)

Unit = 100 km
Length = 2800 km (approx.)

Unit = 50 km
Length = 3400 km (approx.)

Figure 1.13 Measurements of the coastline of Great Britain, the smaller the ruler, the larger the circumference: coastline paradox (Mandelbrot, 1967; the Great Britain map is made by Kevin David Pointon, distributed under a CC0 1.0 license).

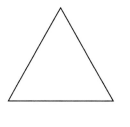 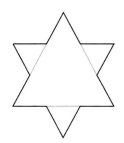

Figure 1.14 the Koch snowflake (distributed under a CC BY 3.0 license).

 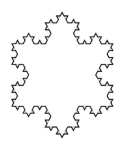

depends on the kind of ruler you use. A smaller ruler will be able to zoom into ever increasing detail of the coast, possible up to every single atom. To make this problem meaningful, we can instead study how length varies as a function of the ruler and see if any systematic variation emerges.

In order to study this effect, we develop it on a geometry that is termed the "Koch snowflake." This is created embedding the shape within itself an infinite number of times, as shown in Figure 1.14. This type of embedding is also better known as a fractal.

How does one quantify the circumference of an ever-increasing complexity? The relationship we are seeking is between the length of the ruler and the circumference. Take first a ruler whose length (L) is equal to the side of the triangle. For that unit length, the circumference is $C = 3$. Now we go to the next figure (Star of David) and decrease L by 1/3, then we find $C = 12$. As we increase the complexity, we find a constant: each time we divide the ruler by 3, the circumference is multiplied by 4. There is a constant value here, which does not vary with the length:

$$D = \frac{\log(4^k)}{\log(3^k)} = \frac{\log(4)}{\log(3)} \tag{1.11}$$

This constant is termed the fractal dimension. Fractals are infinitely complex patterns that are self-similar across different scales. Fractals can be observed in many shapes in nature, see Figure 1.15. The term D is the fractal dimension and quantifies the complexity of the form. So even if a size appears infinitely large, we can still attach a finite number to it – D – with the understanding the length is function of the ruler G used

$$\text{Length}(G) \sim G^{1-D} \quad D > 1 \tag{1.12}$$

Figure 1.15 Self-similar shapes occurring in nature (Romanesco Broccoli created by Ivar Leidus, distributed under a CC BY-SA 4.0 license; Galaxy of Galaxies created by Jonathan J. Dickau, distributed under a CC BY-SA 3.0 license; Milky Way Galaxy created by Nick Risinger, distributed under a CC BY-SA 3.0 license).

What appears now is no longer a logarithm but a power function. Proportional growth (logarithm) becomes fractal growth (power). Do note that the logarithm comes in handy

$$\log(\text{Length}(G)) \sim (1 - D)\ \log(G)\ D > 1 \tag{1.13}$$

Keep in mind the logarithm on both sides; we will use this to our advantage in detecting this type of variation in datasets.

The form $x^{-\alpha}$ is used to define the so-called Pareto distribution:

$$P(X > x) = 1 - F_X(x) = x^{-\alpha} \tag{1.14}$$

where $F_X(x)$ is a cumulative distribution model and α is the shape parameter which determines the steepness of the tail of the Pareto distribution. The name is after a Paris-born (1897) Italian engineer, Vilfredo Pareto, who showed that the distribution of wealth in Europe followed a certain pattern: 80% of wealth lies with 20% of the population. The rich hog their wealth, what else is new?! This 80–20 rule is associated with the $x^{-\alpha}$ form.

1.4.6   Quantile Plots for Extreme Values

In Section 1.4.2, we learnt about the QQ plot, which compares two different distributions. In this section, we introduce quantile plots for extreme values. We compare quantiles from our dataset with theoretical probability distributions models (exponential distribution and Pareto distribution). Quantile plots help us better understand those long tails from extreme value datasets.

The Exponential Quantile Plot

In a similar vein as for the lognormal distribution, we'd like to design a quantile plot that can be used as a diagnostic tool to check whether data have self-similarity, and hence follow a Pareto distribution. To understand how this works, we start first from another type of distribution, the exponential distribution (see also Section 5.2.2). The probability density function is as follows:

$$f_X(x, \lambda) = \lambda \exp(-\lambda x) \tag{1.15}$$

where λ is a shape parameter. The cumulative distribution function has a similar form:

$$P(X \le x) = F_X(x, \lambda) = 1 - \exp(-\lambda x) \tag{1.16}$$

where Fx(x,lamda) is the cumulative distribution function, $P(X \le x)$, the probability that the random variable X is no greater than x.

Note that the cumulative distribution function is calculated from the model of Eq. (1.15). The empirical cumulative function mentioned earlier is calculated empirically from the data. Like the Pareto distribution, this model has only one parameter. An exponent has a nice property: the inverse of an exponent is a log, so we find

$$p = 1 - \exp(-\lambda x) \rightarrow x = -1/\lambda \log(1 - p) \tag{1.17}$$

For the exponential distribution, the quantile function $Q(p)$ (see also Section 5.2.2) has a closed-form expression:

$$Q(p) = -1/\lambda \log(1 - p) \tag{1.18}$$

We can use this equation to calculate the theoretical quantiles of the exponential distribution. We can plug in any p and find the corresponding x. What is noticeable about Eq. (1.18) is that x (the quantile function) is a linear function of $\log(1 - p)$. This is a very helpful property that we will use when we compare the quantiles of the data with the quantiles of the exponential model. To make such a comparison, we do the following

- We rank the data $x_1^* \le x_2^* \le x_3^* \le \dots$; x_1^* is the smallest value in our dataset.
- Each x_i^* is associated with a percentile $p = i/n$ (the probability of being less than x_i^*).
- For that p, we calculate the quantile of the exponential distribution with $\lambda = 1$: $-\log(1 - i/n)$.
- We compare in a plot $(x_i^*, -\log(1 - i/n))$; $i - 1, \dots, n$.

There is slight twist here: when $i = n$, we get the $\log(0)$, which is infinite. Statisticians have come up with a simple trick for this, which they call the continuity correction. As above, it simply means replacing n with $n + 1$. So finally, we get to the exponential quantile plot

$$\left(x_i^*, -\log\left(1 - \frac{i}{n+1}\right) \right), \ i = 1, \dots, n \tag{1.19}$$

The Exponential Quantile Plot for Extremes

The exponential quantile plot we covered so far looks at all values, small and large, and compares them with quantiles of the exponential distribution. What if we want to focus only on the larger values, say above some specified threshold t? We are in luck here as the exponential distribution has a very quaint property: memorylessness.

What is this property? Think of someone providing you with two light bulbs, one burned for 100 hours, the other 150 hours; which one will you pick? Light bulbs have some average lifespan, for example 1000 hours. This doesn't mean that each lightbulb will die after exactly 1000 hours: some will die earlier, some later. This means that the lifespan of a lightbulb has some distribution. This is due to differences that occur in manufacturing bulbs. Some experimenting has shown that the distribution of a lifespan (a random variable) has a form like the exponential, it decreases fast. This means that there are not very many lightbulbs that shine for more than the mean (1000 hours). However, it is also found that if you were given two lightbulbs, each having been used for a different period of time, both would still have the same exponential distribution with the same mean. You would never know which of these two bulbs has been used the most. This property is a direct consequence of the exponential form. Mathematically, this reveals itself by the fact that an integral of an exponential function is still an exponential function. Say, for example, we are interested not in values from 0 to ∞, but starting from some threshold x and going to ∞. In this case,

$$\int_{x}^{\infty} \exp(-y)\mathrm{d}y = \exp(-x) \tag{1.20}$$

and if we take one exponential distribution, that shape is maintained above any threshold x. This is not true for the normal or lognormal distribution. You can use this property to show that in the lifetime of two lightbulbs (represented by the random variables X and Y) the following holds true:

$$P(X > a + 100\,\mathrm{h} \mid X > 100\,\mathrm{h}) = P(Y > a + 150\,\mathrm{h} \mid Y > 150\,\mathrm{h}) \tag{1.21}$$

So, to answer our original question: it does not matter whether you take the 100-hour or the 150-hour bulb, they have no memory!

What this means in terms of quantile plots is that we can focus on data above a threshold and still use the Eq. (1.19). This is shown in Figure 1.16.

For this reason, the expression of the quantile plots for extremes looks slightly different. Instead of ranking from small to large, it is more convenient to rank from large to small:

$$x_1^* \geq x_2^* \geq x_3^* \geq \dots \tag{1.22}$$

This means we need to make one change in Eq. (1.19) and that is to replace i with $n - i$, or

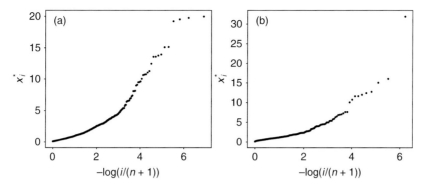

Figure 1.16 Exponential quantile plots of the two diamond datasets: (a) dataset 1, (b) dataset 2.

$$\left(x_i^*, -\log\left(\frac{i}{n+1}\right)\right), \ i = 1, \ldots, n \tag{1.23}$$

The Pareto Quantile Plot

In an exponential quantile plot we are looking for a linear relationship between x and $\log(1-p)$. We will now show that for the Pareto distribution, a linear relationship exists between $\log(x)$ and $\log(1-p)$; we simply need to take an extra log. Note that the Pareto distribution has the form $P(X > x) = 1 - F_X(x) = x^{-\alpha}$, for some threshold x; hence, if we take the logarithm on both sides, then

$$-\log\Big(P(X > x)\Big) = \alpha \log(x) \tag{1.24}$$

If we now take x_i^* (with the largest i) as the threshold x, then

$$-\log\Big(P\big(X > x_i^*\big)\Big) = \alpha \log\big(x_i^*\big) \tag{1.25}$$

The log appears on both sides of the equation, and hence we have a linear relationship between $\log(P(X > x_i^*))$ and $\log\big(x_i^*\big)$. Let's focus on the term $P(X > x_i^*)$. This term is also equal to

$$P\big(X > x_i^*\big) = \frac{i}{n+1} \tag{1.26}$$

simply because we have, in our dataset, i values that are larger than x_i^*. The Pareto quantile plot is as follows:

$$\left(\log\big(x_i^*\big), \ -\log\left(\frac{i}{n+1}\right)\right), \ i = 1, \ldots, n \tag{1.27}$$

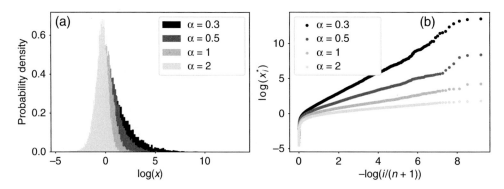

Figure 1.17 (a) Histograms of log Pareto distributions and (b) Pareto quantile plots with different α.

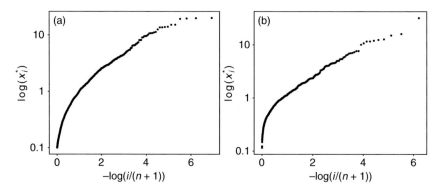

Figure 1.18 Pareto quantile plots of the two diamond datasets: (a) dataset 1, (b) dataset 2.

If this is a straight line, the indication is that our data follow a Pareto model. In addition, the slope of this line is an indication of the value of α. Figure 1.17 shows what the effect is of α, the larger α, the "thinner" the tail of the distribution. A model with a small α means that we are expecting more extremes than a larger α. The slope in the Pareto quantile plot is the reciprocal of $\alpha : 1/\alpha$.

Let's apply this to our two diamond datasets, see Figure 1.18. What do we observe? For one, in both cases we notice that there is not a simple straight-line behavior between the data and the quantiles of the Pareto model. However, for dataset 2, a clearer straight-line behavior emerges for larger values, while, in dataset 1, we seem to level off to a plateau. This difference is an indication of the difference in the way large values behave.

▶   Play **Video 01: Quantile Plots and Extreme Value Index** to learn how to make and interpret Pareto quantile plots.

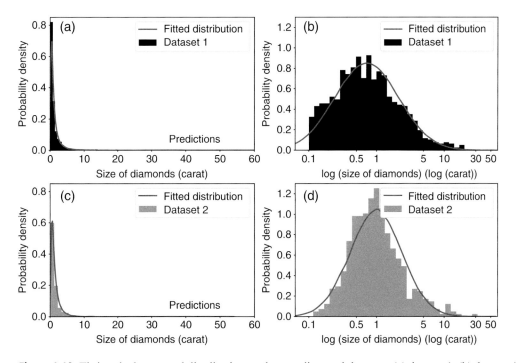

Figure 1.19 Fitting the lognormal distribution to the two diamond datasets: (a) dataset 1; (b) dataset 1 with log *x*-scale; (c) dataset 2; (d) dataset 2 with log *x*-scale.

1.4.7 Fitting Models to Data to Make Predictions

So far, we have analyzed shapes, sizes, and magnitudes in nature and observed that consistent patterns emerge, some associated with a proportional growth, others with infinitely repeating patterns. That's nice, but it does not yet provide what we are really after: making predictions. In particular, we are interested in large values, see Figure 1.19a and 1.19c. To do so, we need to fit a distribution model to data, then use that model to make predictions.

Fitting distribution models to data is a more advanced topic that we will not cover in detail. Note, however, that you do not need to know the detailed derivations, you can simply run functions in software packages that fit distribution models to the data. For most distribution models there aren't any closed-form expressions between the parameters and the data. That is not true for the lognormal distribution, which can be fitted as follows:

- consider a dataset as consisting of values $x_1, x_2, \ldots, x_n$
- take the logarithm of this dataset: we take our data, $y_1, \ldots, y_i, \ldots, y_n$; $y_i = \log(x_i)$
- the parameter μ is estimated by the arithmetic mean:

$$\bar{y} = \frac{1}{n} \sum_{i=1}^{n} y_i \tag{1.28}$$

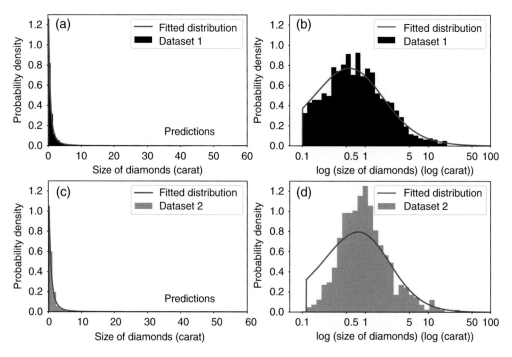

Figure 1.20 Fitting the Pareto distribution to the two diamond datasets: (a) dataset 1; (b) dataset 1 with log *x*-scale; (c) dataset 2; (d) dataset 2 with log *x*-scale.

- the parameter σ is estimated by the empirical standard deviation:

$$s = \frac{1}{n} \sum_{i=1}^{n} (y_i - \bar{y})^2 \tag{1.29}$$

We take our two diamond datasets and fit to both the lognormal and the Pareto distribution models, as shown in Figure 1.19 and Figure 1.20.

Now that we have a model to make a prediction: we'd like to know how many large diamonds (i.e., > 40 ct) would be found in these two deposits. Alternatively, how many stones on average do we need to mine to find one such stone? What's the probability of getting diamonds greater than 40 ct? If we use the lognormal distribution model for both datasets, then the probabilities of getting diamonds greater than 40 carat for two datasets are:

$$P(X > 40 \text{ ct}) = 0.0000905 \ or \ 0.00905\% \text{ for dataset 1}$$

$$P(X > 40 \text{ ct}) = 0.0000124 \ or \ 0.00124\% \text{ for dataset 2}$$

We need to mine $1/P(X > 40 \text{ ct}) = 11{,}055$ or 80,607 stones on average before we find one stone greater than 40 ct. There is a problem with this prediction. Our previous analysis showed that dataset 2 likely has a longer tail, and any model based on that dataset should predict more diamonds of more than 40 ct than dataset 1. What happened? In both predictions we assumed that the dataset follows a lognormal distribution. This is clearly not true for dataset 2, in fact, it is also not true for dataset 1. The latter is clearly visible when looking at the fit for diamonds between 0.1 and 0.5 ct (the small ones).

The problem here is that the lognormal distribution asks for lognormal perfection, where every part of our distribution should be lognormal. In our case, we do not measure anything less than 0.1 ct. So, the distribution maybe lognormal for large values, but not for small values. If we want to predict the distribution only for large values, then why not fit a model to the entire dataset? In the next section we show how developing an extreme value theorem is a win–win solution:

- Win 1: we only need to focus on large samples in the data.
- Win 2: we don't need to choose between models, the data will choose the model.

1.4.8 What Have We Learned in Section 1.4?

- We have learned that variation of sizes and magnitudes may follow a distinct pattern. Since these quantities are positive and because large sizes are less frequent than small, the logarithm appears as a natural mathematical model.
- We have learned that an alternative to the logarithmic model occurs in terms of self-similarity when size may be infinite. Using the fractal dimension, we can make a finite quantification of this possibly infinite size.
- We have learned that, from these patterns, probability distributions emerge, such as the lognormal distribution and the Pareto distribution, which can be used to describe the data using a mathematical model.
- We have learned that QQ plots are ways to investigate what pattern we are dealing with: logarithmic or self-similar.

1.5 Extreme Value Theory

What we are after eventually is to come up with a distribution model that focuses only on large values. Moreover, we would like to use just one model, without having to choose between many different ones, and making assumptions that we don't know are true or will find difficult to test. Extreme value theory provides us with such mathematical models. We will not delve too deeply into this theory, which is an advanced topic. Instead, we will explain why this theory is needed, and show intuitively how it emerges.

1.5.1 Two Ways to Look at Extremes

There are two ways of looking at large values (see Figure 1.21):

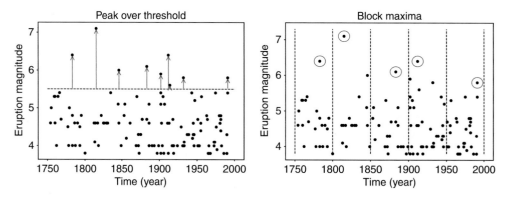

Figure 1.21 Two ways of looking at extremes, with the example of volcanic eruption magnitudes: (a) peak over threshold; (b) block maxima.

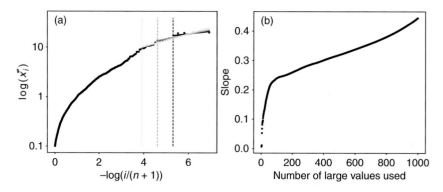

Figure 1.22 (a) Fitting a straight line to the large values of the Pareto quantile plot, while increasing the threshold (i.e., decreasing the number of large values we use). (b) Fitting the slope of the line to the amount of large values used for the diamond dataset 1.

- Peak over threshold: we define a threshold and retain only values over this threshold. This does require defining a threshold.
- Block maxima: we block the data into groups and retain only the maximum value in each group; hence we need to define the size of these blocks.

With block maxima, there is an understanding that grouping samples has some practical meaning. Samples can be grouped, for example, based on time intervals: we wish to understand how maxima vary within a fixed interval. For that reason, we focus more on this topic in Section 1.6, where we consider the frequency of extremes.

We consider peak over threshold by revisiting the Pareto quantile plot; see Figure 1.22. Recall that the Pareto quantile plot can be used to assess whether or not large values follow the Pareto distribution, simply by looking for a straight line between the data and the Pareto quantiles. We can go a step further, and actually fit a straight line to the data, but only using those data above a given threshold. Then, we repeat this, but increasing the threshold each time, which also

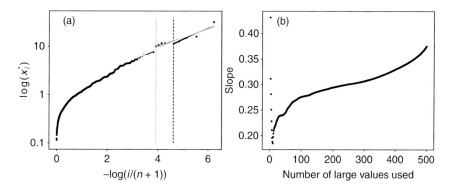

Figure 1.23 (a) Fitting a straight line to the large values of the Pareto quantile plot, while increasing the threshold (i.e., decreasing the number of large values we use). (b) Fitting the slope of the line to the number of large values used for the diamond dataset 2.

means that increasingly fewer data are used. In Figure 1.22b, we plot that slope as a function of the number of large values used. We notice that as we increase the threshold, which also means the number of large values decreases, the slope keeps on decreasing, appearing to converge to zero.

Now, let's move to the second dataset and repeat the same exercise; see Figure 1.23. Now, it appears that the slope remains relatively stable (around 0.2) when we increase the threshold.

What this exercise shows is that we do not need to assume any distribution model to assess the behavior of large values. It appears that one single value, some index, here represented as a slope, indicates the behavior of large values. In the next section, we formalize this as an extreme value theorem.

1.5.2 Extreme Value Theorems

Now let's leverage computational power to understand extreme value theorems through Monte Carlo experiments, without mathematical derivations. We will learn about specific extreme value distributions and see how we can use them on extreme value datasets.

Central Limit Theorem

Theorems in mathematics are essentially statements that have been proven true, but usually only when given certain conditions. Theorems require proofs, which often need to be done algebraically (with symbols and derivations). Such proofs can be quite challenging to understand and outside the scope of this course. Instead, we illustrate this Central Limit Theorum proof using a thought experiment, backed up by a computer simulation.

Thought experiment: imagine you have a sample dataset with billions of diamonds, all from the same deposit. Now put the stones in groups of 10, so we get millions of these groups. For each group, we calculate the average, in carats, so we get millions of averages. We can even make a histogram of them. Consider now that instead of taking groups of 10, we make groups of 100 stones. Again, we take averages, and plot a histogram. It seems reasonable to assume that the average of 10 stones is going to be very similar to the average of 100 stones. What about the spread around this average? When calculating averages of 100 stones, we are more confident in this average than with 10 stones; after all, we have more data ($100 \gg 10$). We expect this average to fluctuate less. What about the shape of this histogram? Here you can imagine, as we increase even more to 1000 or 10 000, that the spread will become even smaller; hence the data will be more and more centered around the mean, so the distribution will become increasingly symmetric.

Computer experiment: we can perform this thought experiment on a computer. Such experiments are termed Monte Carlo experiments (see also Section 5.2.3). An algorithm mimics the sampling that a human would do in the real world by sampling from a theoretical distribution model.

Figures 1.24 and 1.25 present the results of these computer-generated samples from a standard lognormal distribution. When the sample size increases, the spread becomes smaller, the average seems to converge, and the distribution of average is no longer a right-skewed

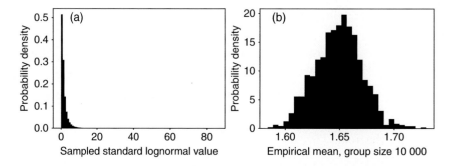

Figure 1.24 (a) Histogram of computer-generated data with a standard lognormal distribution. (b) Histogram of the averages of 10 000 computer-generated data.

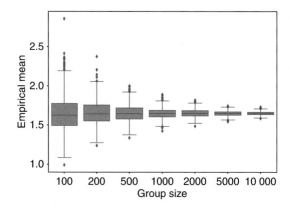

Figure 1.25 Boxplot to indicate the behavior of the average of the standard lognormal distribution as the sample size increases.

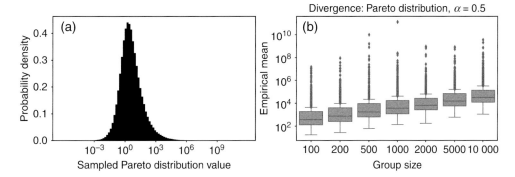

Figure 1.26 (a) Log histogram of computer-generated data from a Pareto distribution ($\alpha = 0.5$). (b) Boxplot showing the behavior of the average of the Pareto distribution ($\alpha = 0.5$) as the sample size increases.

lognormal distribution. In fact, the central limit theorem states that the distribution of average becomes a normal distribution when the sample size gets larger.

However, the central limit theorem does not apply to some Pareto distributions. A Pareto distribution has a fatter tail than a lognormal distribution. Figure 1.26a shows a histogram of a Pareto distribution where $\alpha = 0.5$. The x-axis is in the log scale. Figure 1.26b shows that when the sample size increases, the spread doesn't become smaller, the average seems to diverge and still increase. In fact, Pareto distribution has an infinite mean when $\alpha \leq 1$ and an infinite variance when $\alpha \leq 2$. The central limit theorem assumes the underlying distribution has a finite variance. In the next section, we will learn a new theorem to apply to extreme value distributions.

Distribution Models for Large Values

Now we perform the same thought experiment as before but focusing on the maximum in a sample dataset. This is the first way to look at the extremes in Section 1.5.1, focusing on maxima.

Thought experiment: imagine the same sample with billions of diamonds, all from the same deposit. Again, we put these stones in groups of 10, so we get millions of these groups. For each group, we now only retain the maximum values in a sample. We can even make a histogram of these maxima. Consider now that instead of making groups of 10, we make groups of 100 stones, again we take the maxima, and plot a histogram. Logically, this histogram of maxima from 100 stones has higher values than using only 10 stones, so we standardize (see Section 5.3.1) the maxima as follows using the mean ($\overline{maximum}$) and standard deviation ($s_{maximum}$):

$$maximum \leftarrow \frac{maximum - \overline{maximum}}{s_{maximum}} \tag{1.30}$$

What would the histogram of the maxima, blocked into samples of 10 and blocked into samples of 100, look like? Here, we turn to a computer experiment. In this experiment, we perform this exercise for two different distributions: the exponential distribution and the normal distribution (Figure 1.27). We do this to check if we get different results for each distribution.

Figure 1.27 Plots showing the computer-generated data sampled from the theoretical normal and exponential distribution.

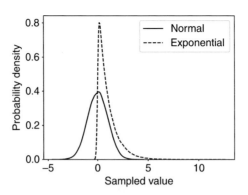

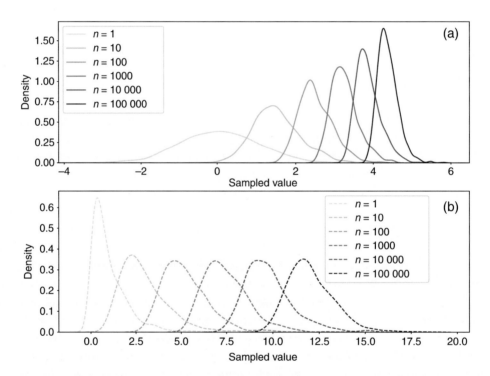

Figure 1.28 Distribution of the maximum values as the sample size increases: (a) normal distribution; (b) exponential distribution.

Turning to Figures 1.28 and 1.29, we observe that regardless of the type of distribution, the distribution of the maxima is always the same. So, it looks like we are discovering that the very high values behave very similarly, regardless of the distribution type. This is a powerful observation, because it suggests that in order to estimate extremes we do not need to know what type of distribution model we are dealing with.

Now we return to the peak over threshold method; see Figure 1.30 and 1.31. Here we repeat the exact same experiment, but instead of choosing the maximum values, we choose all values

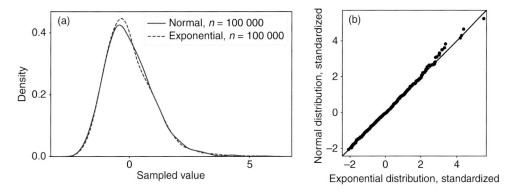

Figure 1.29 (a) Distribution of the normal and exponential maximum for n = 100 000, but standardized. (b) QQ plot between the standardized exponential maximum and the normal maximum. Note that there is no difference if the model is exponential or Gaussian.

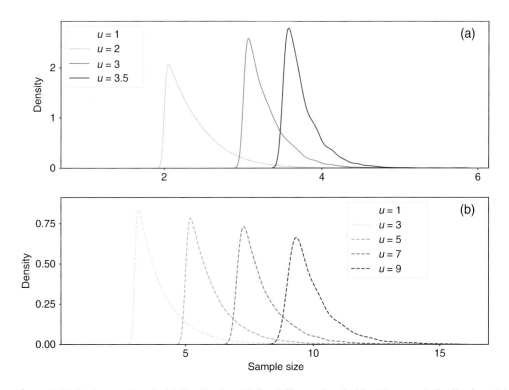

Figure 1.30 Peak over threshold distribution (u) for different thresholds: (a) normal distribution; (b) exponential distribution.

above a threshold. Clearly, as the threshold increases, these values become increasingly larger, so again we perform a standardization.

The result in Figure 1.31 shows that, like with maximum values, the distribution of peaks over threshold (termed "u") is the same regardless of what type of distribution you have.

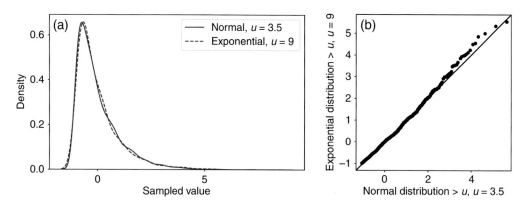

Figure 1.31 (a) Standardized peak over threshold distribution; (b) QQ plot of the standardized exponential and normal peak over threshold.

Extreme Value Distributions and Theorems

Now we are ready to state some formal theorems. The detail of these theorems is outside the scope of this textbook, so we just provide them pro forma. There are two theorems, one for maxima, the other for the peaks over threshold. These theorems are found algebraically (with symbols and derivations), not computationally, but the algebra follows the same reasoning as for the computer experiments.

For maxima, we find that, regardless of the type of distribution, the maximum follows the generalized extreme value (GEV) distribution:

$$P\Big(\max(X_1,\ldots,X_n)\leq x\Big) = GEV(x) = \begin{cases} \exp\left(-\Big(1+\xi\big(\tfrac{x-\mu}{\sigma}\big)\Big)^{-1/\xi}\right), \xi\neq 0 \\ \exp\left(-\exp\big(-\tfrac{x-\mu}{\sigma}\big)\right), \xi=0 \end{cases} \tag{1.31}$$

Let's unpack this equation a bit. We are looking for a distribution function, not of data, but of the maxima of data, hence the $\max(X_1,\ldots X_n)$. We are also defining a cumulative distribution function: the probability P of a maximum value less than or equal to x. This distribution has three parameters, a center parameter μ, a scaling parameter σ, and a very special parameter ξ termed the extreme value index. When this index ξ equals 0, then we find that this GEV distribution becomes a double exponential distribution $\exp(-\exp(-x))$ (see Figure 1.32).

The extreme value index is the key parameter here, it affects the shape of the function $GEV(x)$, the μ and σ do not affect the shape. Because the index ξ is so important, various cases have historically been given different names:

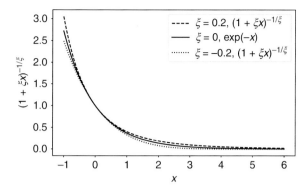

Figure 1.32 A plot showing that $(1 + \xi x)^{-1/\xi}$ becomes $\exp(-x)$ when $\xi \to 0$.

- $\xi < 0$, the Weibull distribution: this distribution has an upper bound, meaning it stops at some large value
- $\xi = 0$, the Gumbel distribution: a type of exponential behavior
- $\xi > 0$, the Fréchet distribution: a type of behavior with tails "longer" or "heavier" than the exponential type

What does the equation mean? It means simply that if we know ξ, μ, and σ, we know the probability distribution of the maximum values. Knowing the probability distribution allows us to make predictions of the frequency of occurrence of the maxima.

For peak over threshold, the values larger than a threshold u follow the generalized Pareto distribution (GPD):

$$P(X \le x | X > u) = GPD(x) = \begin{cases} 1 - \left(1 + \xi\left(\dfrac{x - u}{\sigma}\right)\right)^{-1/\xi}, \xi \ne 0 \\ 1 - \exp\left(-\dfrac{x - u}{\sigma}\right), \xi = 0 \end{cases} \tag{1.32}$$

The ξ is the same one as for the maxima, σ is a scaling parameter. Note we divide them into three families

- $\xi < 0$: distributions have an upper bound: they have a maximum that is finite
- $\xi = 0$: a large family of distributions with a moderate tail (e.g., exponential)
- $\xi > 0$: a family of distributions with a very long tail (e.g., Pareto).

Estimating the Extreme Value Index ξ

The statistical literature provides several ways of estimating ξ, and some of this material is quite technical. Here, we provide an intuitive way using quantile plots. Recall that the Pareto quantile plot can be used to identify ξ, simply by looking at the slope or the 1/slope of the quantile plot for large values. The limitation here is that this only works for $\xi > 0$. Indeed, $\xi > 0$ in the GPD contains the Pareto distribution as a special case.

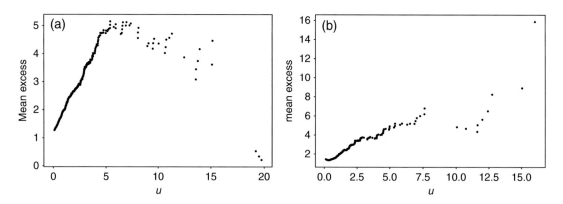

Figure 1.33 Mean excess plots for both diamond datasets: (a) dataset 1; (b) dataset 2.

We need a quantile plot that we can use to identify any value of ξ, negative, zero, or positive. We are in luck. Statisticians have found that when plotting the Pareto quantile plot of the "mean excess," again a straight-line behavior emerges, now including all values of ξ. What is this mean excess? It is the average of values over a given threshold, with the threshold then being subtracted; in math terms:

- take a threshold $u = x_k^*$
- retain all data x_i^* larger than u: $x_1^*, \ldots, x_{k-1}^*$
- calculate the mean of the retained data: $\dfrac{1}{k-1} \sum_{i=1}^{k-1} x_i^*$
- subtract u from that mean: $\dfrac{1}{k-1} \sum_{i=1}^{k-1} x_i^* - u$

Figure 1.33 shows how this mean varies with the threshold u. The mean excess becomes noisy at high thresholds u due to too few samples.

In the next step, we check how this mean varies in a Pareto quantile plot; hence we plot

$$\left(-\log\left(\frac{k}{n+1} \right), \ \log\left(\frac{1}{k-1} \sum_{i=1}^{k-1} x_i^* - x_k^* \right) \right), \ k = 1, \ldots, n \tag{1.33}$$

The excess is the amount over a threshold, what is left over above u, because we subtract u. And each time we calculate a mean.

Figure 1.34 shows the mean excess quantile plot for the two diamond datasets. An interesting pattern emerges: for dataset 2 we again observe that the mean excess quantile plot has a linear trend upward, while for dataset 1 that does not happen. There is more good news. Mathematical statisticians have proven that the line you fit is again a good estimate of ξ. The novel thing about the mean excess quantile plot is that, unlike the Pareto quantile plot, we can use it to identify all cases of ξ.

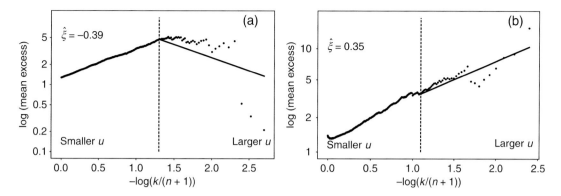

Figure 1.34 Mean excess quantile plots for both diamond datasets: (a) dataset 1; (b) dataset 2.

The final question in estimating ξ is to find a suitable threshold u. There are various ways of doing this with sophisticated statistics, but often finding a good point where the mean excess quantile plot stabilizes, as shown in Figure 1.34, is a good start. What should you be looking for? You need to know what the consequences are when you take the threshold too high or too low:

- too high: you get very few values over the threshold, so the fitting of the slope becomes unreliable (we call this "variance")
- too low: we are now including too many small values; hence, we are not estimating properly the extremal behavior of the data (we call this "bias").

Next to estimating ξ, we also need to estimate the center parameter u and scale parameter σ in Eq. (1.32). Again, we refer you to software packages to do that for you, the derivations are of little consequence to the material covered here.

How Certain Are We about the Estimated ξ?

This question is very important. Recall that one of the uses of data science for the geosciences is to use data to uncover something about the geoscientific process that created the data. For diamonds, lognormal distributions indicate that significant sorting of diamond sizes took place, while Pareto-type behavior suggests that there was no such sorting. This idea therefore can be used to investigate whether the data come from an area close to or far away from the diamond source, the kimberlite volcanoes. The question therefore is: are the data lognormal or not, and perhaps to what degree? This brings us to a very important element of data science: statistical hypothesis testing. Even though the technique is statistical, we want to emphasize that we should not lose track of the original scientific question:

- H_0 (i.e., the null hypothesis): we are far from the source
- the opposite of H_0: we are close to the source.

Here we have a case of a single hypothesis test (a statement and its compliment). Now we make the next logical step, to introduce what is termed in mathematics as a necessary condition. Being far from the source indicates lognormal behavior

- H_0: the data are lognormal
- the opposite of H_0: the data are not lognormal.

It is important to understand, however, that we should not equate "lognormal" with "far from the source." Lognormality of the data is only one necessary condition; it is not in itself sufficient, you may need other conditions, so it is indicative, but not necessarily conclusive. Now, we go yet one step further and create another necessary condition:

- $H_0 : \xi = 0$
- the opposite of $H_0 : \xi \neq 0$.

Again, we have a necessary but not sufficient condition. When $\xi = 0$ we may also have an exponential model. So why do we do it this way? What we have done is to translate a geoscientific hypothesis into a statistical hypothesis that involves one single parameter!

What we need to know now is if $\xi = 0$. The quantile plots provided us already with significant indications; we just need to look for the linear behavior for large values. Sometimes this linear behavior is very clear and the slope steep, for example when $\hat{\xi} > 1$. Sometimes, it not at all that clear, and perhaps $\hat{\xi}$ is close to zero. We need to investigate this quantitatively and not just qualitatively. We need to know what a plausible variation (variance) of $\hat{\xi}$ could be. Note that the hat of ξ is the estimate $\hat{\xi}$ from the slope.

We mentioned the notion of "variance" when estimating ξ based on a few large values. This is why we need uncertainty on estimates. With few data, the estimate $\hat{\xi}$ is likely different from the true ξ. How confident are we about this estimate? Can we definitely exclude the case $\xi = 0$? Is the estimated value close to zero really close enough?

To answer these questions, we need not just one single estimate of ξ but a number of alternative estimates. In Section 5.2.3, we go in greater detail in the procedures used to do so. This procedure is termed "bootstrap" method. Regardless of the procedure, the outcome is a set of alternative estimates of ξ, which you can put into a histogram; see Figure 1.35.

Figure 1.35 shows the estimation of ξ for dataset 1 is less than 0. Therefore, we can reject the null hypothesis H_0. Dataset 1 has a thinner tail and decreases faster than the lognormal. The estimation of ξ for dataset 2 is larger than 0, where the distribution of ξ is centered around 0.35. We can also reject the null hypothesis and conclude that dataset 2 has a longer tail than the lognormal distribution.

▶   Play **Video 01: Quantile Plots and Extreme Value Index** to learn about the extreme value theorem and extreme value index.

1.5.3 Protocol for Predicting Magnitudes of Extreme Events

At various instances, we will provide protocols for applying a certain tool, technique, or methodology. Just as chemists in a laboratory need to follow protocols, data scientists also require certain protocols. These are important for double checking your work; you should never blindly trust the output of a code as you may have loaded the wrong dataset! For that reason, it is important to make all the necessary plots that help you diagnose any problem that may occur. Box 1.1 summarizes the protocol for predicting extreme events.

Box 1.1 Protocol for predicting extreme events

- Calculate the lognormal, Pareto, and mean excess quantile plots
- Use quantile plots to diagnose threshold
- Estimate ξ using regression over threshold
- Run an uncertainty analysis of ξ using the bootstrap method
- Final estimate of ξ, σ using the GPD fitting routine
- Make predictions of extreme events using the fitted GPD

Plots to make

- Histogram of data with summary statistics
- Lognormal, Pareto, and mean excess quantile plots
- Estimation of ξ as function of the threshold
- Histogram of bootstrap estimates of ξ
- Histogram of data overlayed with the fitted GPD

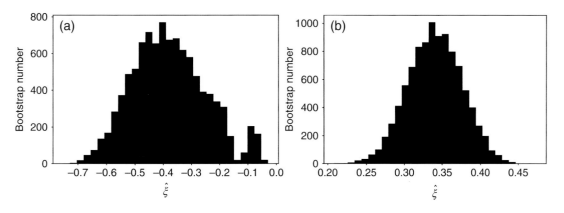

Figure 1.35 Histogram of the estimates of ξ using the bootstrap method: (a) dataset 1; (b) dataset 2.

1.5.4 Application to Diamond Datasets

Now we are ready to apply all this theory to real cases. We will ask ourselves a very practical question. Diamond mining companies know that much of their profit comes from mining a few very large stones. So, we ask ourselves the question, for both deposits: What is the probability of observing a stone larger than 40 ct? Note that this threshold is much larger than the largest stone in our sample.

Here are some additional questions and suggestions for executing the protocol listed in Box 1.1:

- Make a histogram of the data, including basic summary statistics.
- Make a lognormal quantile plot, a Pareto quantile plot, and a mean excess quantile plot.
- Use these diagnostic tools to get insight into the dataset:
  - Are small values behaving differently from large ones?
  - What happens for large values in these quantile plots?
  - Can you identify a threshold above which a linear pattern emerges?
- If the Pareto quantile plot shows a clear linear behavior for large values:
  - guestimate a suitable threshold for extremes
  - estimate ξ by fitting a line
  - quantify the uncertainty on ξ using the bootstrap method
- If the Pareto quantile plot leaves ambiguity, use the mean excess quantile plot to:
  - guestimate a suitable threshold for extremes
  - estimate ξ by fitting a line
  - quantify the uncertainty on ξ using the bootstrap method
  - fit the final ξ and σ using a routine for fitting the GPD
- If ξ is negative, calculate the upper bound (maximum possible value).
- Use the fitted GPD to predict the probability for a diamond to be over 40 ct.

 Visit **Notebook 03: Extreme Value Theory** to follow the protocol and answer the practice question (probability larger than 40 ct).

1.5.5 What Have We Learned in Section 1.5?

- We have learned that two equivalent ways of looking at extremes exist: considering only values over a given threshold and looking at maximum values within blocked sequences.
- We have learned that both these views result in theoretical models that share the same important parameter: the extreme value index.
- We have learned that the magnitude of the extreme value index indicates how "long" or "heavy" the tail of the distribution is.

1.6   Frequency of Extremes

In previous sections, when considering the diamond datasets, we did not need to deal with the concept of time. All stones were put in one parcel, and we disregarded when they were collected, simply because it does not matter. This is not the case when dealing with temporal geoscientific data such as volcano eruptions, earthquakes, or extreme weather events. Now we need to bear in mind not only the size/magnitude but also the frequency in time.

1.6.1   The Return Period

A very common way to quantify this frequency is through a return period. Even in day-to-day usage people talk about the "100-year earthquake" (in a particular area), meaning the size of an earthquake that returns "on average" once in 100 years. On average is important here: it may in fact not occur in the next 100 years, or it could occur twice, but certainly not 10 times! We return to the precise formulation of this in Section 1.6.3.

There is an alternative way of making the same statement. Imagine the magnitude of the 100-year earthquake is 7.5 on the Richter scale. Then the following statement reflects the same information: a magnitude 7.5 or stronger earthquake happens each year with a probability of 1%.

Here we use the volcanic eruption dataset (Figure 1.36) from the beginning of this chapter to visualize the return period. We can put all of the data now into a single plot, which is the return-level plot, shown in Figure 1.37, which presents the volcanic eruption dataset by means of the return level and return period. Let's take volcanic eruption events which have a VEI greater than 6. How many eruptions of this scale happened from 1750 to 2000? There have been five such events. Therefore, the empirical return period T of the return level VEI = 6 is (2000 − 1750)/5 = 50 years. Then, we say that the VEI = 6 event is a "50-year volcanic eruption." We calculate empirical return periods for every return level.

> ▶   Play **Video 02: Return Level and Return Period** to understand the return level and return period of volcanic eruptions.

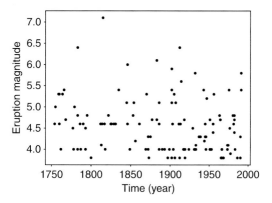

Figure 1.36 Volcanic eruption dataset from 1750 to 2000.

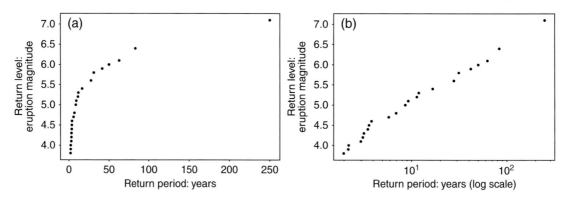

Figure 1.37 Plots of return levels versus return period (a) in years (a) and (b) using a log scale for the return period.

1.6.2   An Extreme Value Model for the Return Period

What is nice about Figure 1.37 is that there is a linear relationship between the log of the return period and the magnitude (which is also on a log scale by definition). If this were true, then we have a means to estimate return periods larger than the observations in the data; see Figure 1.38.

Is it always linear? The answer lies again in the extreme value theorem. Let us consider that we have a case where $\xi = 0$. Then the GPD from Eq. (1.32) is as follows:

$$P(X > x | X > u) \sim \exp\left(-\frac{x - u}{\sigma}\right) \tag{1.34}$$

Hence, when we take the logarithm of $P(X > x | X > u)$ we get a linear form:

$$\log(P(X > x | X > u)) \sim -\frac{x - u}{\sigma} \tag{1.35}$$

We also define a new function, the return period as the inverse of $P(X > x)$:

$$T(x) = \frac{1}{P(X > x)} = \frac{1}{P(X > x | X > u)P(X > u)} \tag{1.36}$$

Why do we do it this way? Say the probability of exceeding a threshold of magnitude 5 equals 0.01, then $1/0.01 = 100$. That magnitude 5 returns, on average, every 100 years. That is why we call $T(x)$ the return period. The magnitude x is the return level. The denominator $P(X > x)$ is represented by a conditional probability $P(X > x | X > u)$, multiplying a peak over threshold u probability $P(X > u)$. More formally,

$$\log\left(\frac{1}{T(x)}\right) \sim -\frac{x - u}{\sigma} \rightarrow \log\left(T(x)\right) \sim \frac{x - u}{\sigma} \tag{1.37}$$

which provides a linear relationship. We now extend this case to when ξ is not zero, which is shown in Figure 1.39. We calculate the logarithm of return period as a function of the return

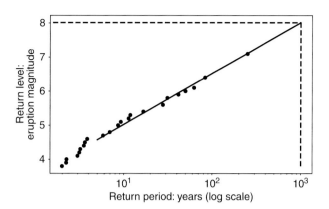

Figure 1.38 Extrapolating large return periods on a return-level plot.

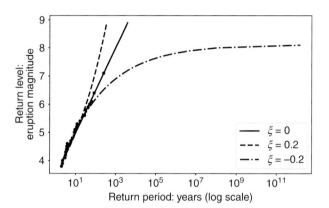

Figure 1.39 Shapes of the theoretical return-level model for various values of ξ.

level using Eq. (1.32). Hence, once we know ξ, which we have learned to estimate in the previous sections, we also have a theoretical return-level model.

1.6.3 Protocol for Predicting the Magnitude and Frequency of Extreme Events

Box 1.2 summarizes the protocol for predicting return periods of extreme events.

1.6.4 Application to the Volcano Dataset

Now we are ready to apply the extreme value theory (Notebook 04) to answer the foillowing:

- "2000-year eruption": What is the size of a volcanic eruption that returns on average once in 2000 years?
- "VEI $\geq$ 8": How long will it take on average to have one VEI $\geq$ 8 eruption?

Note that these two questions have not been observed in this dataset: our time span is 250 years and there is no eruption larger than 8.

Box 1.2 Protocol for predicting frequency of extreme events

- Make a plot of time versus magnitude for various thresholds of magnitude
- Estimate of ξ, σ using the GPD fitting routine
- Make prediction of extreme events using the fitted GPD
- If $\xi < 0$, add the maximum magnitude to the return-level plot

Plots to make
- Magnitude versus time of all events
- Time-blocking plots if looking at maxima
- Exponential quantile plots
- Histogram of bootstrap estimates of ξ
- Fitted return levels versus empirical return levels, including confidence intervals

---

 Visit **Notebook 04: Frequency of Extremes** to follow the protocol and predict the return level of 2000 years and the return period of VEI = 8.

1.7   What Have We Learned in This Chapter?

- We have learned that predicting extremes requires specialized models; that common distribution models often fall short in treating extremes comprehensively.
- We have learned that magnitudes of events often follow certain types of distribution models, such as lognormal and Pareto distributions, which reflect the way events evolve in a physical sense.
- We have learned the importance of the quantile plots to visualize and diagnose how extremes behave in a dataset.
- We have learned practical protocols that, when followed rigorously, allow a confident prediction of the future occurrence of extreme events.

TEST YOUR KNOWLEDGE

1.1 What is the logarithm associated with?
   a. Linear growth
   b. Proportional growth
   c. Geometric mean
   d. Self-similarity

1.2 What elements of a dataset can be diagnosed with a QQ plot?

    a. How that dataset compares to another dataset

    b. How that dataset correlates to another dataset

    c. How that dataset compares with a theoretical model

1.3 Put these distributions in the order of "longest tail" to "shortest tail":

    a. Lognormal

    b. Pareto

    c. Exponential

    d. Normal

    e. Distribution that has an upper bound

1.4 What does the extreme value index measure?

    a. How long the tail of a distribution is

    b. The type of extreme value distribution

    c. The variance of sample values above a threshold

    d. The mean of a distribution

1.5 Why do we need extreme value theory?

    a. It provides theoretical models for the occurrence of extremes

    b. It simplifies the problem of predicting extremes to one important index

    c. Because there is no longer a need to choose a specific distribution model

1.6 What is the return-level plot used for?

    a. To predict the probability of occurrence of an extreme over a threshold

    b. To predict the frequency of extremes

    c. To check how well the fitted extreme value model fits the data

    d. To determine the exponential quantiles of a dataset

1.7 Why do common distribution models such as normal and lognormal often fall short in treating extremes?

    a. They are more difficult to fit to data

    b. They impose a fixed distribution model for the entire range of data values

    c. They do not converge according to the central limit theorem

1.8 In extreme value statistics, we need to decide on the threshold above which we retain large values. Which of the following is true?

    a. If the threshold is too low, the predictions will be biased

    b. If the threshold is too low, the predictions will be very uncertain

    c. If the threshold is too high, the predictions will be biased

    d. If the threshold is too high, the predictions will be very uncertain

1.9 Why do we need to do an uncertainty analysis for the extreme value index

    a. To be able to test the hypothesis of whether $\xi = 0$

    b. To find the best estimate of ξ

    c. To know for sure that $\xi = 0$

1.10 What does $\xi < 0$ entail?

    a. It means that the normal distribution model is excluded

    b. It means that the distribution has an upper bound that can be estimated

    c. It means that quantile plots are not reliable any more

1.11 Why do we use the bootstrap method to get an uncertainty quantification of the estimate of ξ?

    a. To double check the estimation method

    b. To test the hypothesis $\xi = 0$.

    c. To quantify uncertainty in the prediction of the occurrence of extremes

    d. To test whether the dataset is biased

FURTHER READING

Easy reading in extreme value analysis does not really exists, but two good books are:

- Beirlant, J., Goegebeur, Y., Segers, J., and Teugels, J. L. (2004). *Statistics of Extremes: Theory and Applications*, John Wiley & Sons.
- Coles, S., Bawa, J., Trenner, L., and Dorazio, P. (2001). *An Introduction to Statistical Modeling of Extreme Values*, Springer.

Chapter 1 of Beirlant et al. has a lot on quantile plots and how these can be used to diagnose extremal behavior. The rest of the book is more focused on theory development. In Coles et al., we recommend in particular chapters 3 (pp. 45–52), 4 (pp. 74–81), and 7 (pp. 124–130).

REFERENCES

Deligne, N. I., Coles, S. G., and Sparks, R. S. J. (2010). Recurrence rates of large explosive volcanic eruptions. *Journal of Geophysical Research: Solid Earth*, 115(B6). https://doi.org/10.1029/2009JB006554.

Mandelbrot, B. (1967). How long is the coast of Britain? Statistical self-similarity and fractional dimension. *Science*, 156(3775), 636–638. https://doi.org/10.1126/science.156.3775.636

Newhall, C. G. and Self, S. (1982). The volcanic explosivity index (VEI) an estimate of explosive magnitude for historical volcanism. *Journal of Geophysical Research: Oceans*, 87(C2), 1231–1238. https://doi.org/10.1029/JC087iC02p01231.

Siebert, L., Simkin, T., and Kimberly, P. (2010). *Volcanoes of the World: Third Edition*. University of California Press. www.jstor.org/stable/10.1525/j.ctt1pnqdx.

2 Multivariate Analysis

Expected Learning Outcomes

- You will learn that datasets that are compositions require a different treatment than typical multivariate datasets.
- You will learn to make various log-ratio transformations on compositional data and use them to perform exploratory data analysis.
- You will learn about biplots and principal component analysis; both methods allow the visualizion of high-dimensional multivariate and compositional data in lower dimensions thereby allowing for better understanding of these types of data.
- You will learn to perform unsupervised clustering that allows detecting natural grouping in datasets. You will learn to link these groups to natural processes that have taken place.
- You will learn about factor analysis that may aid you in understanding the processes that created variation present in the data.
- You will learn to apply rigorous protocols and practical software that allows studying and predicting with multivariate and compositional data on real datasets.

2.1 Introduction

Geochemical analysis is used for a large variety of purposes. For example, in the geological sciences, this type of analysis is used to understand Earth processes. In sedimentary geology, provenance analysis is used to study the composition of sandstones. The composition of these sandstones reflects the nature of the source rocks exposed, as well as the climatic processes by which sands were generated from these rocks. A major contribution here is the Dickenson Model (Dickinson and Suczek, 1979), which quantifies empirically (by looking at compositional data) the plate-tectonic setting of the sedimentary basin. What does this mean and why is it important? The discovery of plate tectonics is fairly recent. In 1912, Alfred Wegener (a meteorologist) postulated the concept of "continental drift." The hypothesis was that there once existed a single continent (now termed Pangea) and that various "continents" then drifted apart. The theory was not well received, mostly because of a lack of data. That first piece of data arrived as geophysical evidence, namely of paleomagnetism. The fact that rocks of different ages show different magnetization was linked to the moving continents. A second piece of data emerged from studying the ocean sea floor. The spreading of the floor can be linked to changes in magnetic fields. All of these data came about in the 1950s and 1960s, many scientists contributed to developing the theory of plate tectonics from data.

What is key here is that people use data and link them to geological or other Earth processes. This theme was touched upon in the previous chapter. The data are the end product of a process. It is a bit like detective work. You need to use observations (today) to decipher what happened in the past. The data we will be dealing with in this chapter are geochemical and/or compositional in nature. In statistical terms they are often termed "multivariate" (there are multiple variables). It simply means that our sample has not just one value or measurement but many measurements. Often, these values form a "composition." Namely, you take a mass or a volume and analyze the proportion of "stuff" in it. In geochemistry, this is typically the proportion of elements, oxides, sulfides, etc. present. By studying the variation of the composition in multiple samples, we may be able to discover that one group of samples appears very different in composition than another group, indicating that they formed under different circumstances (processes). This may be easy if we have only two or three different elements; it becomes much harder when you have 50 elements. Hence, the chemical interaction between these may be quite complex. Then, the true detective work starts!

Such detective work can be useful for scientific discovery but may also have very practical application. This is what we will cover next.

2.2   Motivating Example: Groundwater Pollution in the Central Valley of California

California's Central Valley is one of the most productive agricultural regions in the world, see Figure 2.1. Agricultural irrigation heavily draws on the groundwater system. At the same time, with an increase in population, groundwater consumption is expected to increase. The result is that pumping from increasingly deeper parts of the aquifer has increased the rate of downward groundwater flow. These increased gradients have been linked to the release of, for example, uranium, an element that is naturally occurring in many different rocks, from sedimentary to volcanic. The question therefore concerns how we can maintain groundwater quality while dealing with this increased need for it. Understanding this trade-off is key to sustainable groundwater management. Simply increasing groundwater by supply, for example through a process of recharge (e.g., flooding a field), may affect its quality. It may lead to an increased introduction of contaminants such as pesticides. In other words, groundwater management actions may have unintended consequences. It is important to understand how groundwater quality is affected by any management actions on, for example, seawater intrusion, land subsidence, or declining water levels. A very good basic text to understand the issues at hand is found in Fakhreddine et al. (2019) upon which our discussion is based.

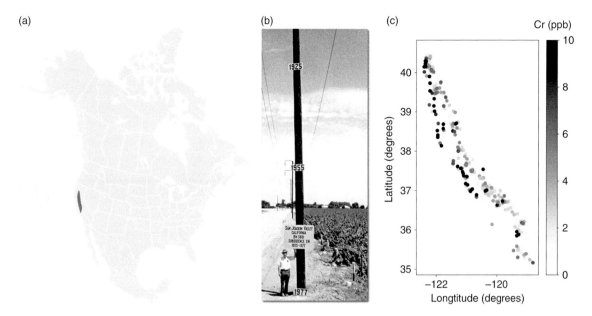

Figure 2.1 (a) Location of the Central Valley of California (California Central Valley Grasslands Map, distributed under a CC BY-SA 3.0 license; Wikimedia Commons, 2010). (b) Subsidence due to overpumping of the groundwater system. The signs on the pole show the approximate altitude of the land surface in 1925, 1955, and 1977. The site is in the San Joaquin Valley southwest of Mendota, California (photographed by Joseph F. Poland, US Geological Survey) (Poland, 1977). (c) Groundwater concentration (parts per billion [ppb]) of chromium (Cr) from January 2018 to January 2019. A black and white version of this figure will appear in some formats. For the color version, refer to the plate section.

To understand the interplay between quality and quantity enhancements, one needs to understand the fundamental physical and chemical processes that are taking place.

"Physical" here means groundwater flow and general hydrology. In this chapter, we focus on the chemical part. Because of the pervasive contamination issues in the Central Valley, a substantial amount of geochemical analysis has been collected from existing wells. We wish to understand the important processes, either natural or anthropogenic, that cause variation in these data. We may find a certain "signature" or "patterns" in the data that can determine the process of contamination in a particular area. This will help decision making in terms of taking the right kind of actions: a very challenging problem because many processes can cause statistical variation. For example, the Central Valley has what are termed "geogenic" contaminants, which means it has arsenic (As), chromium (Cr), uranium, which you don't want to drink, naturally occurring. A simple analysis could look for high levels of these elements, indicating possible anthropogenic contamination, although they may be naturally occurring. The point we wish to make is that a signature of a feature is more than a single high value. What we are looking for is a combination of elements and of a certain composition.

We will focus on chromium as a contaminant. Hence, a bit more chemistry knowledge is needed as a context.

(a)   (b)

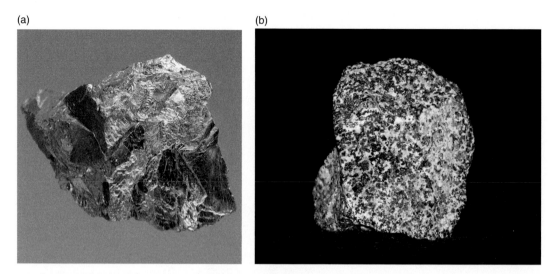

Figure 2.2 (a) Pure chromium (Chromium, distributed under the CC BY-SA 3.0 license; Wikipedia Commons (2008). (b) Chromite ore (Chromite, distributed under the CC BY-SA 4.0 license; Wikipedia Commons (2009). A black and white version of this figure will appear in some formats. For the color version, refer to the plate section.

2.2.1 Chromium as a Contaminant

Chromium from the Greek "chroma," meaning color, is a hard and lustrous metal, mostly known for its anticorrosive properties. It is popular because it can be polished shiny without tarnishing. Chromium alloys are produced from a mineral called chromite; see Figure 2.2.

Ferrochromium alloy is commercially produced from chromite by silicothermic or aluminothermic reactions, and chromium metal by roasting and leaching processes followed by reduction with carbon and then aluminum. Chromium metal is of high value for its high corrosion resistance and hardness. A major development in steel production was the discovery that steel could be made highly resistant to corrosion and discoloration by adding metallic chromium to form stainless steel. Stainless steel and chrome plating (electroplating with chromium) together comprise 85% of the commercial use.

In the United States, trivalent chromium, the Cr(III) ion, has been considered an essential nutrient in humans for insulin, sugar, and lipid metabolism. While chromium metal and Cr(III) ions are not considered toxic, hexavalent chromium, Cr(VI), is both toxic and carcinogenic. Abandoned chromium production sites often require environmental clean-up. Hexavalent chromium is much more mobile in groundwater than trivalent chromium because it does not bind easily to minerals within soils and sediments.

Trivalent chromium, under certain conditions will transform into hexavalent chromium. Industrial activities are important sources of hexavalent chromium contamination in California, particularly in the San Francisco Bay and Los Angeles areas. However, naturally occurring chromium can be related to the occurrence of very common types of rock: mafic and ultramafic. Soils derived from mafic/ultramafic parent material contain high amounts of chromium, leading to above-average chromium concentrations in soils. Therefore, the process whereby hexavalent chromium exists in the environment may have many sources. Geochemist state there are many "pathways" for it to get into the environment. What we would like to do is to unravel some of those pathways from data.

2.2.2 Water Quality Data for the California Central Valley

 Visit **Notebook 05: Central Valley Groundwater Geochemistry data** to download California Central Valley water quality data and perform some exploratory data analysis.

The questions we will address are:

- What combination of elements are indicative of a human impact in water quality versus a natural occurrence?
- What caused this impact? Agriculture? Pollution?
- Where in the Central Valley can we find these combinations of elements, thereby informing mitigation action?

2.3 Overview of a Typical Statistical Analysis for Geochemistry

First, we will list a general protocol of how any analysis could proceed, then we provide details on each step.

Step 1: What questions are you interested in addressing? A statistical analysis should have a goal, which drives the way you do the analysis. The goal can be about "understanding" the processes going on that shape your data, or the goal can be to make "prediction," possibly of what could be happening at unmeasured locations.

Step 2: Exploratory and visual analysis of compositional data. Getting insight into datasets where many measurements are taken in one sample, termed multivariate data, is challenging, in particular when you have more than five measurements. A typical geochemical analysis may have up to 100 measurements per sample. In addition, we are dealing with special kinds of multivariate data, termed compositional data: in each sample everything sums up to 100%, we are studying parts of a whole. We show that this "closed sum" makes regular statistical analysis biased, and hence we need to do a transformation on the data, to remove this issue. For compositional data, specialized statistical summaries and plots need to be introduced. At the end of this step, you will be able to get insight into what elements in the samples are correlated, or what elements you could concentrate on for further analysis.

Step 3: Outlier detection and, possibly, removal. An outlier is different from an extreme because it is an anomaly, it is not part of the population, at least from a statistical point of view. Outliers need to be addressed, and explained. Either they are very interesting, and we can start by focusing on outlier samples, or they are due to some measurement error. Detecting them is challenging because we cannot simply look at each element in a sample separately, we need to consider the whole. Therefore, we need to use methods that can detect outliers in multivariate data.

Step 4: Find communality in the data by clustering. Samples that "look the same" are likely, but not necessarily, to result from the same common process. Hence, we prefer to group samples that are very similar to each other. This grouping is termed clustering in statistics. The first problem is that we do not know how many groups exist, unless experts who know more about the chemical processes of the data can help with this. Secondly, the grouping needs to be based on the entire sample, not on individual elements: again, we need to look at the whole, not just the parts. Once we have done the grouping, we can also plot the group labels on a map and observe whether or not labels plot together in space. This provides us with spatial insight into the grouping, possibly allowing us to discover something about why those labels plot together.

Step 5: Finding factors that determine the composition of the samples. Usually, a few major factors influence the composition of our sample; for example, the location being in an agricultural area versus an urban area. In factor analysis, we try to determine how many factors there are and what they are. It may be much more complex than simply an urban versus agricultural area; hence, a detailed factor analysis, supported with insight into the processes that may have taken place, can provide such determination.

These steps can be applied in a geochemical analysis, and also potentially in any multivariate analysis.

2.4   Concept Review

In this chapter, we move to higher dimensions, two, three, and even higher dimensions that are hard to visualize. In Section 5.3, we provide a refresher on some basic statistical measures for multiple random variables as well as some important geometric properties of higher dimensions. More specifically, if you need such refresher, familiarize yourself with the following:

- Standardizing multiple variables such that they have a common unitless scale.
- An alternative to the correlation coefficient as measures of correlation, namely the covariance.
- Geometric properties of ellipses and ellipsoids. Knowing these will help you understand and visualize correlation in higher dimensions.
- Eigenvalue decomposition of symmetric matrices. In multivariate analysis, we will deal frequently with symmetric matrices. Their properties can be visualized using ellipses and ellipsoids.

2.5   Compositional Data Analysis

Geologists measure the composition of minerals in rocks. Nutritionists look at the proportions of types of food in a diet. Sociologists look at the racial distribution of a population. All of these cases are examples of compositional data. This means that many types of data are

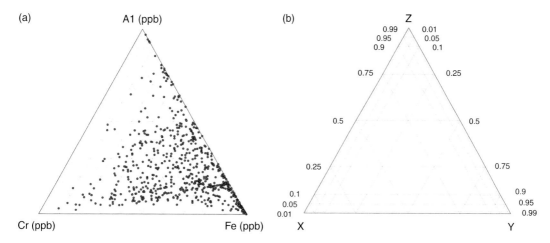

Figure 2.3 (a) Using ternary diagrams to plot compositional data. The square is the center of the compositional dataset. (b) Reading a ternary diagram in terms of composition.

compositional in nature. What is common here is that we are studying "parts" of some "whole," a total. In other words, there is a constant sum involved. In many cases the sum is 100%. We do this because, if geologists only studied the "mass" of minerals, with different samples having different masses, the difference in the mass in minerals would also be due to the difference in the mass of the samples. In other words: we are mostly, if not only, interested in relative differences. Compositional data can be visualized using ternary diagrams, where we can plot three parts of the whole at one time. This is illustrated in Figure 2.3.

 Using CoDaPack, make ternary diagrams for the Central Valley compositional dataset. Please refer to CoDaPack_tutorial.pdf.

Compositional data are also multivariate in nature. The term multivariate is statistical nomenclature for stating that we are not dealing with just one variable at time (the size of a diamond), but many variables (e.g., the size and value of a diamond). What differentiates compositional and multivariate data from univariate data is the notion of correlation.

2.5.1 Quantifying Correlation in Multivariate Datasets

When two variables are associated, it is often an indication of a process that generated such an association. Hence, quantifying "association" may lead to the discovery of the process that caused the association. We would like to emphasize the word "may": as we point out later, blindly interpreting correlation as causation is fraught with difficulties.

An association between two quantities refers to a broad concept and description, without necessarily a stringent mathematical definition. Linear correlation, often abbreviated as

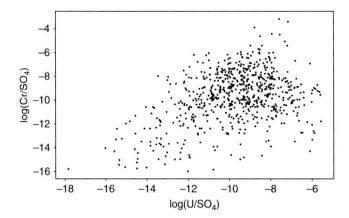

Figure 2.4 A scatter plot as a visual way to appreciate correlation: the logarithm of the ratio between the uranium concentration and sulfate concentration versus the logarithm of the ratio between the chromium concentration and sulfate concentration.

"correlation," refers to the narrow concept that two quantities have a linear relationship. When dealing with only two variables, the simplest way to visualize association is by a scatter plot; see Figure 2.4.

Besides visual confirmation, we need a quantitative measure of the degree of correlation. This value is termed the (linear) correlation coefficient. Consider a dataset of samples of two variables: $(\mathbf{x}, \mathbf{y})$

$$(x_1, y_1), (x_2, y_2), \ldots, (x_n, y_n) \tag{2.1}$$

Then the correlation coefficient is defined as

$$r = \frac{\sum_{i=1}^{n}(x_i - \bar{x})(y_i - \bar{y})}{\sqrt{\sum_{i=1}^{n}(x_i - \bar{x})^2 \sum_{i=1}^{n}(y_i - \bar{y})^2}} \tag{2.2}$$

We provide some visual appreciation in Figure 2.5. Notice that this coefficient is between $r = -1$ (perfect anticorrelation) and $r = +1$ (perfect correlation), and that no linear correlation means that r is zero.

The value of r should be interpreted with care. For instance, in Figure 2.6, r measures only the linear correlation, it is not a measure for non-linear association.

Figure 2.7 illustrates that r is very sensitive to outliers, and hence it is important to detect outliers. Finally, Figure 2.8 shows that correlation does not mean causation.

2.5.2 Limitations of Applying Correlation to Compositional Data

There is a fourth limitation to the use of correlation and that is when we deal with compositional data. A simple example of such dataset is one with three samples, and we are studying four components (matter) contained in each sample: animal, vegetable, mineral, water

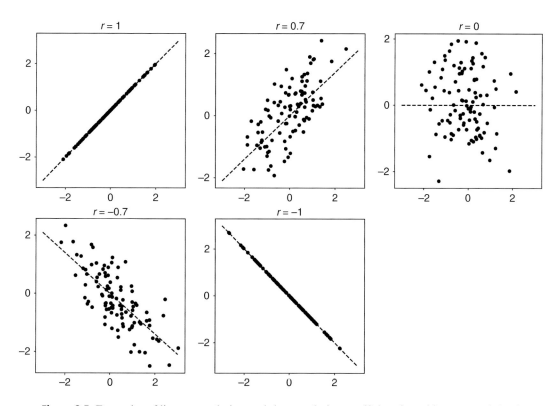

Figure 2.5 Examples of linear correlation and the correlation coefficient for arbitrary sampled values.

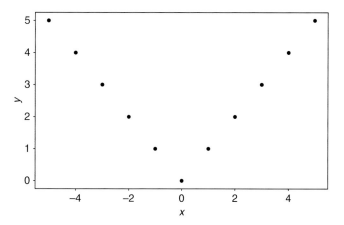

Figure 2.6 Non-linear relationship between y and $\log x$: $r = 0$ and does not measure any non-linear relationship.

(Table 2.1). We can use Eq. (2.2) (where x and y are two of the four components) to calculate the correlation coefficients, which are listed in Table 2.2. When we have more than two quantities or variables, we typically plot correlation coefficients in matrix form. On the diagonal we find "1" simply because a quantity is perfectly correlated with itself.

Now we remove one component: water, for example, when we dry the sample (Table 2.3). Practically, this means that we recalculate the composition (proportions) with only three, as

Table 2.1 The proportion of four components of matter in three samples of a compositional dataset

	Animal	Vegetable	Mineral	Water
Sample 1	0.1	0.2	0.1	0.6
Sample 2	0.2	0.1	0.1	0.6
Sample 3	0.3	0.3	0.2	0.2

Table 2.2 The matrix of pairwise correlation coefficients

	Animal	Vegetable	Mineral	Water
Animal	1	0.5	0.87	−0.87
Vegetable	0.5	1	0.87	−0.87
Mineral	0.87	0.87	1	−1
Water	−0.87	−0.87	−1	1

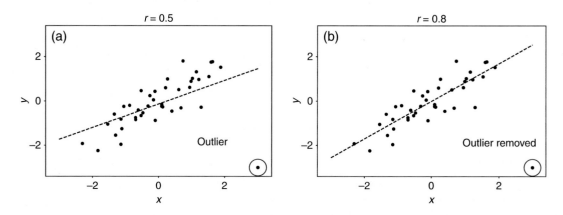

Figure 2.7 When plotting y versus x, r is sensitive to outliers: (a) r, keeping outlier (b) r, removing outlier.

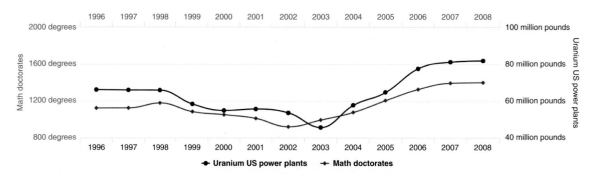

Figure 2.8 The correlation of math doctorates awarded with uranium stored at US power plants does not entail causation (Vigen, 2015, distributed under CC BY 4.0).

Table 2.3 Compositional of the three samples after drying

	Animal	Vegetable	Mineral
Sample 1	0.25	0.5	0.25
Sample 2	0.5	0.25	0.25
Sample 3	0.375	0.375	0.25

Table 2.4 The matrix of pairwise correlation, after drying all samples

	Animal	Vegetatable
Animal	1	−1
Vegeable	−1	1

listed in Table 2.4. Note here that the mineral variable does not vary so we leave it out. We now notice that the correlation coefficients have changed, in fact now we find that "animal" and "vegetable" matter are perfectly negative correlated. This doesn't make any sense because the only thing we did was remove the water.

Why is this? To understand it, we have to consider that compositions have a closed sum. For example:

$$X + Y + Z = 1 \tag{2.3}$$

is a composition, but that also means that you can write one component as a function of the others:

$$X = 1 - Y - Z \tag{2.4}$$

So, regardless of what the data are, X is always a function of Y and Z, hence X will always be correlated to Y and Z, even if in the real world this correlation is meaningless or simply doesn't exist. Imagine randomly adding some "salt" to each sample. There is no correlation between "salt" and the rest because we randomly added it; yet

$$Salt = 1 - X - Y - Z \tag{2.5}$$

So, if we know X, Y, and Z, we will also know exactly the amount of salt, despite the fact there is no correlation between X, Y, Z, and salt! Luckily, there is a way to fix this problem, and Figure 2.4 already gives a clue, namely, instead of correlation between X, Y, and Z, we look at the logarithm of ratios.

2.5.3 Key Requirements in the Analysis of Compositional Data

When studying correlation in compositions, we have to be a little more careful than when dealing with non-compositional data. Here we state a few logical principles that should emerge from how we treat such data.

Scale invariance: the analysis should not depend on how much you sum up to. This sum in compositional data analysis is also termed the "closure constant." Here is a simple example (Pawlowsky-Glahn et al., 2011): consider you are studying a sandstone. This sandstone erodes and becomes sand, deposited elsewhere. You would like to find out how the composition of the original sandstone changes as it becomes sand. Suppose you measure the composition as Q = quartz, F = feldspar, O = other minerals or rock fragments.

In the sandstone we find: $[Q, F, O] = [55, 40, 5]$ %
In the sand we find more quartz $[Q, F, O] = [80, 15, 5]$ %

Interpretation 1: If we have 100 g of sandstone, it will contain 55 g of quartz. If you have 55 g of quartz in the sand, that would mean that the total sample weight would equal 68.8 g. Hence, in that sample, we would then also have 10.3 g of feldspar and 3.4 g of other. Conclusion: 29.7 g of feldspar and 1.6 g of other have been removed.

Interpretation 2: If we have 100 g of sandstone, it will contain 40 g of feldspar. If you have 40 g of feldspar in the sand, the total sample weight would equal 266.7 g, hence in that sample we would then also have 213.3 g of quartz and 13.3 g of other. Conclusion: 158.3 g of quartz and 8.3 g of other have been added.

Clearly, compositions do not convey information about mass, and if you considered the mass instead you would reach conflicting interpretations. Compositions only convey relative information.

Permutation invariance: a second requirement is that the analysis does not change according to how you permute the variables in a composition; for example, the conclusions you draw should not change if you analyze $[Q, F, O]$, $[F, Q, O]$, $[O, Q, F]$, etc.

Subcompositional coherence: a third requirement is that the conclusions you reach should not change when you make subcompositions. In the above example in Tables 2.2 and 2.4, we indeed saw such change; hence, we need a new framework that deals with all three issues. That framework is the "log ratio." Figure 2.9 shows correlations under log-ratio transformations are the same when we make subcompositions.

2.5.4 Log Ratios

We have seen those log ratios help us preserve the subcompositional coherence. But how are we going to choose the base of the log ratios? In Figure 2.9, we use the mineral as the base. Are there any other options? Do we have other types of log ratios that are helpful? Next, we provide three different options for log ratios: additive, centered, and isometric, and consider their summary statistics.

The Additive Log Ratio

The first concept we introduce is the idea of working with ratios. Take a simple example

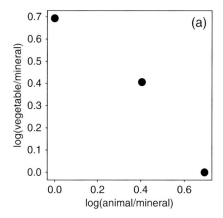 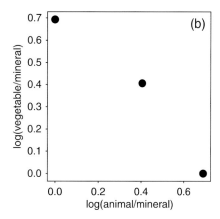

Figure 2.9 Subcompositional coherence of log ratios in terms of correlation coefficients: (a) log-ratio correlation before water removal ($r = -0.98$); (b) log-ratio correlation after water removal ($r = -0.98$).

- the probability that it rains tomorrow is 80%
- the probability that it rains the day after tomorrow is 20%

What is the difference between 80 and 20, how would you say this correctly with "words"? You can't say: there is 60% more probability! The simplest way is to say: it is four times more likely to rain tomorrow than the day after tomorrow. The number 4 comes from dividing 80 by 20 or taking a ratio. Let's explore this further with more parts, for example a composition of four parts:

$$20, 30, 40, 10$$

We can take the ratio with respect to the last component:

$$2, 3, 4, 1$$

One problem with ratios is that they are not symmetric:

$$20/10 \neq 10/20$$

This can be an issue because obviously our analysis should not depend on whether we do 20/10 or 10/20: both convey the same information. We have pointed out already several times that we want to work with properties that are "invariant." A simple solution is to take the logarithm of the ratio:

$$[\log(2), \ \log(3), \log(4), \log(1)]$$

Note that:

$$\log(10/20) = -\log(20/10)$$

So, they are equal up to a sign, which is fine because we can always square things:

$$\Big(\log(10/20)\Big)^2 = \Big(\log(20/10)\Big)^2$$

Another nice property of the log ratio is that it can be any value, it does not need to be between 0 and 1. There are no bounds or closed sums for log ratios. We can always add one log ratio to another one. This is not true for compositions; you cannot simply keep adding probabilities, eventually they will become larger than one!

Now we are ready to define our first log-ratio method. Consider that

$$\mathbf{x} = [x_1, x_2, \ldots, x_D] \tag{2.6}$$

is a D-part composition, with $x_i > 0$, $i = 1, \ldots, D$. Then we define the additive log-ratio (alr) transformation as

$$\mathbf{y} = \mathrm{alr}(\mathbf{x}) = \left[\log\left(\frac{x_1}{x_D}\right), \ldots, \log\left(\frac{x_{D-1}}{x_D}\right)\right] \tag{2.7}$$

Hence, in the additive log ratio, we make ratios with respect to one part. This transformation is bijective, which simply means that you can calculate $\mathbf{x}$ from $\mathbf{y}$ and $\mathbf{y}$ from $\mathbf{x}$:

$$\mathbf{x} = \mathrm{alr}^{-1}(\mathbf{y})\text{: } x_i = \frac{\exp(y_i)}{\exp(y_1) + \ldots + \exp(y_{D-1}) + 1} \tag{2.8}$$

A very striking example of the benefit of using log ratios is shown in Figure 2.10. Figure 2.10a shows a scatter plot of chromium (Cr) versus uranium (U) ion concentration (in ppb) of our samples. It looks rather disappointing if we had hoped to see some correlation. If we take the same dataset, but we take the ratio of the metal ions to the sulfate (SO_4) anion, we can see a clearer relationship (Figure 2.10b). Geochemists may in fact point out that taking such ratio makes a lot of sense, since $U(SO_4)_2$ is a salt that can dissolve in water, and chromium(III) sulfate, $Cr_2(SO_4)_3$ is also a solute.

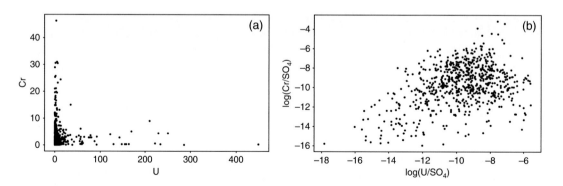

Figure 2.10 (a) Scatter plot of U vs Cr. (b) Scatter plot of the log ratio of U/SO_4 vs Cr/SO_4.

> 🖳 Visit **Notebook 05: Central Valley Groundwater Geochemistry data** to do log-ratio scatter plots for other compositions.

The Centered Log Ratio

The disadvantage of the additive log ratio is that we have to choose what x_D is, so the formulation is not quite symmetric. Therefore, an alternative can be formulated by making the ratio with the geometric mean $g(\mathbf{x})$. This ratio is termed the centered log ratio (clr):

$$\mathbf{z} = \text{clr}(\mathbf{x}) = \left[\log\left(\frac{x_1}{g(\mathbf{x})}\right), \ldots, \log\left(\frac{x_D}{g(\mathbf{x})}\right) \right], \quad g(\mathbf{x}) = \left(\prod_{i=1}^{D} x_i \right)^{\frac{1}{D}} \tag{2.9}$$

and with back transformation

$$x_i = \frac{\exp(z_i)}{\exp(z_1) + \ldots + \exp(z_D)} \tag{2.10}$$

The disadvantage of this ratio is that we still have a closed sum:

$$\sum_{i=1}^{D} \log\left(\frac{x_i}{g(\mathbf{x})}\right) = 0 \tag{2.11}$$

> 🖳 Visit **Notebook 05: Central Valley Groundwater Geochemistry Data** to do clr transformations and compare with alr transformations.

The Isometric Log Ratio

A third log ratio has been formulated that we will not discuss in detail here because of the advanced mathematics it involves, and that is the isometric log ratio (ilr). You will see it appear in software. This ratio does not have the disadvantages of the alr or clr: the ilr is symmetric and does not have a closed sum. But it is more difficult to interpret. You can refer to Egozcue et al. (2003) for the mathematical details.

Treatment of Zeros

An important issue to address is what to do when the sample value or one of the compositions is zero. Obviously, we cannot take the logarithm of zero. The answer is that it depends. Here are three different situations with three different solutions:

Case 1: the part with zeros is not important for the study. Solution: the part should be omitted.

Case 2: the part is important: the zeros are essential. We divide the sample into two populations according to the presence/absence of zeros. Then we study each population. This is common practice too when dealing with datasets that have positive values and zeros. The zeros are treated by separating the dataset in two parts.

Case 3: the part is important, but the zeros are due to rounding. Devices have limited detection limits, hence anything below the limit appears to be zero. In that case, we could for example replace the zero with the detection limit itself.

Statistical Summaries of Log Ratios: Standardizing Log-Ratio Data

Standardizing univariate data (see Section 5.3.1) is achieved by taking each sample x, subtracting the mean, x_1 and dividing by the standard deviation σ

$$x \leftarrow \frac{x - \bar{x}}{\sigma} \tag{2.12}$$

The result of such standardization is that the mean of the data equals zero and the standard deviation equals one. This allows us to compare different datasets, for example, which have different units (kilograms vs meters). When dealing with multivariate data, then we can repeat this for each variable. The center of multivariate data is the mean of each variable:

$$\bar{\mathbf{x}} = (\bar{x_1}, \bar{x_2}, \ldots, \bar{x_D}) \tag{2.13}$$

In standardizing compositional data, we cannot use the same idea, simply because we cannot use plus/minus operations or multiplication/division. Why is this? Take two compositions with three components, and add them up:

$$\mathbf{x} + \mathbf{y} = (x_1, x_2, x_3) + (y_1, y_2, y_3) = (z_1, z_2, z_3) = \mathbf{z} \tag{2.14}$$

The z is no longer a composition. The same problem will occur when doing divisions with a scaler. We need different ways of doing additions and multiplication. The route followed is inspired again by the logarithm:

$$\log(xy) = \log(x) + \log(y) \tag{2.15}$$

In other words, "summing" is now similar to "multiplying." For that reason, we use a new type of sum, $\oplus$, defined as follows:

$$\mathbf{x} \oplus \mathbf{y} = \left(\frac{x_1 y_1}{x_1 y_1 + x_2 y_2 + x_3 y_3}, \frac{x_2 y_2}{x_1 y_1 + x_2 y_2 + x_3 y_3}, \frac{x_3 y_3}{x_1 y_1 + x_2 y_2 + x_3 y_3} \right) = (z_1, z_2, z_3) = \mathbf{z} \tag{2.16}$$

We note again the division by the total to retain the compositional nature of $\mathbf{z}$. Likewise, we have

$$\log(x^c) = c \log(x) \tag{2.17}$$

This means we can use the "power" $\odot$ to multiply a composition with a constant c,

$$c \odot \mathbf{x} = \left(\frac{(x_1)^c}{(x_1)^c + (x_2)^c + (x_3)^c}, \frac{(x_2)^c}{(x_1)^c + (x_2)^c + (x_3)^c}, \frac{(x_3)^c}{(x_1)^c + (x_2)^c + (x_3)^c} \right) \tag{2.18}$$

Having this knowledge, we can construct a method to standardize compositional data.

First, we need to think carefully about what the "center" is. We have already learned in Chapter 1 that using the arithmetic mean is not a good idea with problems that involve logarithms, so it is better to use the geometric mean. Because we have D parts in our composition, we can calculate the geometric mean of each part:

$$g_i = \left(\prod_{j=1}^{n} x_{ij} \right)^{\frac{1}{n}} \tag{2.19}$$

where x_{ij} is the j-th sample value of part i. This means that instead of one geometric mean, we have as many such means as we have parts in the data. How do we use these various g_i to define a single center like we did with multivariate data? The answer is to use the barycenter, which is also termed the "closed geometric mean" and calculated as follows. First, we consider the $\widetilde{g}_i$ as a composition itself but make sure the sum equals 1:

$$\widetilde{g}_i = \frac{g_i}{\sum_{i=1}^{D} g_i} \tag{2.20}$$

Then the barycenter has coordinates:

$$\mathbf{g} = (\widetilde{g}_1, \ldots, \widetilde{g}_D) \tag{2.21}$$

Once we have a center, we can "subtract" it from the data, which we now know how to do:

$$\mathbf{x} \leftarrow x \oplus \mathbf{g} \tag{2.22}$$

Figure 2.11 shows an example of what this operation achieves. Clearly, the effect of centering is that the small blip of data in Figure 2.11a is now centered in the middle. This also means that isolines that define the ternary diagram have shifted as well. The square is the center of the data.

 Using CoDaPack, make ternary diagrams and center ternary diagrams. Please refer to CoDaPack_tutorial.pdf.

If we look at Figure 2.11b, we are still disappointed that, while the blip of data is in the center, it is still a small blip. We'd like to scale it. To do so we need to calculate a measure of total standard deviation, v_{tot}. If we have such measure, we can simply do:

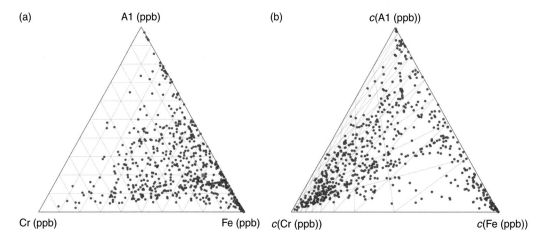

(a) A1 (ppb)

Cr (ppb)   Fe (ppb)

(b) c(A1 (ppb))

c(Cr (ppb)) c(Fe (ppb))

Figure 2.11 Effect of centering on a ternary diagram: (a) ternary diagram, the square is the center; (b) ternary diagram after centering. Now the center of the data is located at the middle of the ternary diagram.

$$\mathbf{x} \leftarrow \frac{1}{\sqrt{v_{tot}}} \odot (x \oplus \mathbf{g}) \tag{2.23}$$

We start with calculating the variance of the log-ratio data (with n samples and D part compositions)

$$v_{ij} = \mathrm{var}\left(\log\left(X_i/X_j\right)\right) = \frac{1}{n}\sum_{k=1}^{n}\left(\log\left(\frac{x_{ik}}{x_{jk}}\right) - m\right)^2 \tag{2.24}$$

with m the following mean

$$m = \frac{1}{n}\sum_{k=1}^{n}\log\left(\frac{x_{ik}}{x_{jk}}\right) \tag{2.25}$$

Because we have D values in our composition, we can calculate all possible v_{ij} and arrange these into a matrix V of size $D \times D$:

$$V = \begin{pmatrix} v_{11} & v_{12} & \cdots & v_{1D} \\ v_{21} & v_{22} & \cdots & v_{2D} \\ \vdots & \vdots & \ddots & \vdots \\ v_{D1} & v_{D2} & \cdots & v_{DD} \end{pmatrix} \tag{2.26}$$

A measure of the total variance is simply obtained by summing all elements in this matrix:

$$v_{tot} = \frac{1}{2D}\sum_{i=1}^{D}\sum_{j=1}^{D}v_{ij} \tag{2.27}$$

Table 2.5 An example of a variation array for the Central Valley dataset output by CoDaPack

X_i/X_j	Cr (ppb)	NO$_3$ (ppb)	SO$_4$ (ppb)	U (ppb)	Ba (ppb)	Ca (ppb)	clr variances
			Variance ln (X_i/X_j)				
Cr (ppb)		3.0060	4.7090	5.3227	2.8771	2.4694	1.5573
NO$_3$ (ppb)	6.7447		5.7725	4.3911	4.1130	3.5544	1.9662
SO$_4$ (ppb)	9.6910	2.9464		3.9455	3.8088	1.7179	1.8189
U (ppb)	−0.0369	−6.7816	−9.7280		4.4961	3.1504	2.0443
Ba (ppb)	3.9651	−2.7796	−5.7259	4.0021		0.9066	1.1936
Ca (ppb)	10.1734	3.4287	0.4824	10.2104	6.2083		0.4598
Mean ln (X_i/X_j)							9.0401	X_i/X_j

Statistical Summaries of Log Ratios: The Variation Array

In the variation array, we combine mean values of log ratios with variances of log ratios into one single table or matrix. In the lower triangle we store the mean values; in the upper triangle, the variance. This is shown in Table 2.5, for a subcomposition of the Central Valley dataset. We notice also some coloring provided for low and high variance. Low variance means that there two elements are likely co-varying because their ratios have low variation (e.g., barium [Ba] and calcium [Ca]), while high variances are often of interest, meaning there is significant variation between these two elements, for example Cr and SO$_4$.

 Using CoDaPack, make a variation array and think about the explanation for it. Please refer to CoDaPack_tutorial.pdf.

2.5.5 Visualizing Compositional Data in a Biplot

Compositional data are multivariate data. A scatter plot on all dimensions of compositional data is not easy to interpret. In the next sections, we introduce the visualization tool on plotting multivariate/compositional data in two-dimensions.

Geometric Interpretation of Correlation

Before addressing how we can visualize compositional data in plots, let us revisit "regular" data. Let's start simple with something in two dimensions and revisit the scatter plot shown in Figure 2.12.

We take the scatter plot and center the data, creating the black axes. Then we rotate the scatter plot such that the new x-axis explains the maximum variance. An ellipse allows us to visualize this. What we end up is a scatter plot that shows no correlation at all. We could go even further and stretch the ellipse to get a circle. Then we are in a new situation

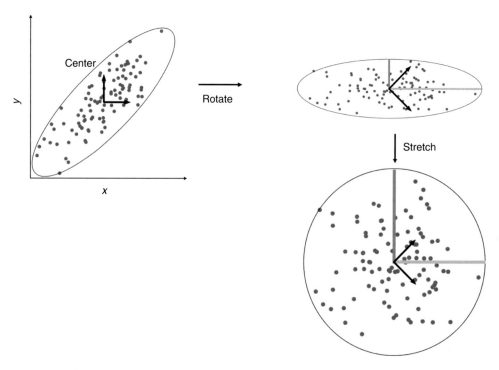

Figure 2.12 Geometric interpretation and transformation of a scatter plot.

where both variables have the same scale and are not correlated. Notice also that, as we rotate and stretch, we create new lines or axes, the dark- and light-grey lines (green and orange, respectively, if you are viewing the electronic version), which define a new axes system. In this new system, we can also evaluate what the contribution is of the black axes because they are now rotated. We also notice that unlike the black axes, the grey axes have different lengths. This length is important. For example, if dark grey is equal to light grey after the rotation, there is no correlation in the original data. If the dark grey axes are very small, while the light grey are very large, then the data are very correlated. In the following sections, we call the new axes lines (dark grey and light grey) the principal component vectors.

Visualizing Multivariate Data in a Two-Dimensional Plot

Let's now focus on extending this idea to three instead of two dimensions. To do this we extend the idea of having an ellipse (in two dimensions) to an ellipsoid (in three dimensions). Now things get more complicated since we need to do a bit more than a single rotation to get rid of correlations. To understand this, let us conceptualize the problem with a football; see Figure 2.13. If you describe the position and geometry of the football, idealized as an ellipsoid, then you would need nine values:

(a)

(b)

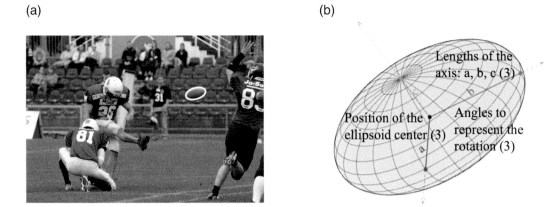

Figure 2.13 (a) Football as an ellipsoid. (b) Nine values to describe an ellipsoid (*American Football Kick*, photographed by Torsten Bolten, distributed under a CC BY-SA 3.0 license; *Ellipsoid*, distributed under a CC BY-SA 4.0 license).

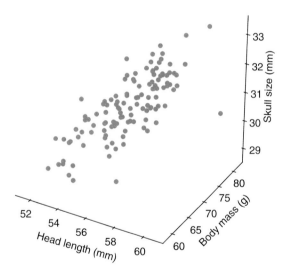

Figure 2.14 Trivariate dataset: the scatter plot looks like an ellipsoid.

- three for position of the ellipsoid center
- three for the lengths of the axis
- three angles that represent the rotation

Why is this relevant? Figure 2.14 shows an example of trivariate data: body measurements for samples of blue jays (Tarvin, 2002). We notice that the scatter of the data follows this simple ellipsoid structure. The task is to represent these data by a two-dimensional plot. To do this, as shown in Figure 2.15, we first find the two major axes of the scatter of point. Then ask how they are rotated. Once we have done that, we can position ourselves looking down on the third axis. In mathematics, we term this a "projection," which means we squash it all into

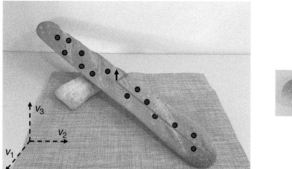

Figure 2.15 Representing a three-dimensional ellipsoid in two dimensions, inspecting (a) Three-dimensional ellipsoid data (b) two-dimensioal plot of the three-dimenional ellipsoid data.

Figure 2.16 Pairwise scatter plot of a trivariate blue jay dataset. On the diagonal they are the density plots of each variable. Off the diagonal they are the scatter plots.

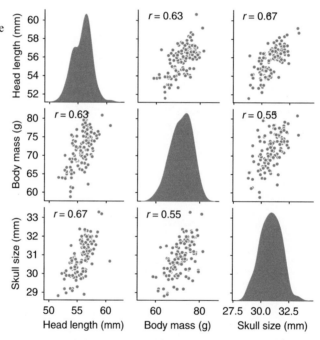

a two-dimensional plot. As we are doing this, we also plot the axes that represent the value or variables we are studying, as well as the sample data.

Let's now apply this idea of projection to two datasets. The first one is the blue jay dataset with three variables. We can plot the matrix of pairwise scatter plots as shown in Figure 2.16. The head length, body mass, and skull size of blue jays are correlated. A larger blue jay will have a longer head, more weight, and a larger skull size. If now we make the projection in two dimensions, we get the plot in Figure 2.17. This plot is called a biplot. We notice that the head length and skull size variables are aligned, their axes coincide in this plot, while the body mass appears orthogonal. What this shows is that when the arrows of each variable are aligned, they

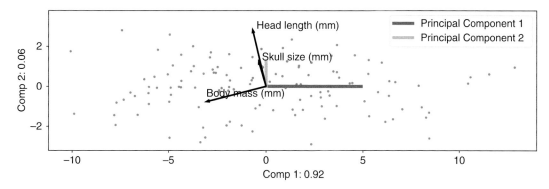

Figure 2.17 Biplot of a trivariate blue jay dataset. Biplot axes are corresponding to the principal components we mention in Section 2.7. The *x*-label and *y*-label have the percentage of explained variance for each component: 92% and 6%.

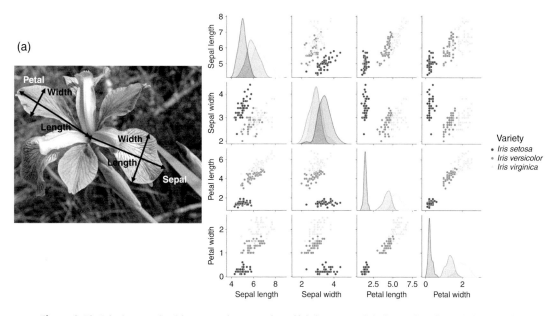

Figure 2.18 Iris dataset, looking at various species of iris in terms of their petal and sepal characteristics, such as width and length. (a) Photo of an iris (*Iris Virginica*, photographed by Frank Mayfield, distributed under a CC BY-SA 2.0 license) (b) Pairwise scatter plot of petal and sepal characteristics.

are highly correlated in this two-dimensional projection; when they are orthogonal, they are not correlated.

Let's now move to a more complex dataset with four variables; see Figure 2.18, which is the famous iris dataset studies by the founder of statistics, Ronald Fisher (Dheeru and Casey, 2017).

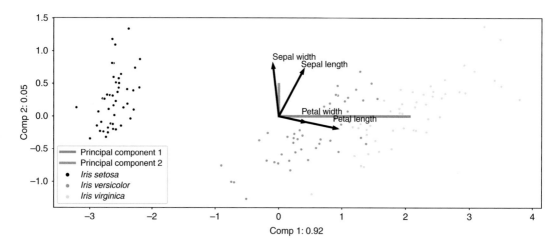

Figure 2.19 Biplot of the iris dataset.

Figure 2.19 shows the biplot of the iris dataset. What should you look for? Recall that arrows indicate the properties you are studying while the dots are the sample (the various iris flowers).

- Look for alignment between arrows, which indicates correlation: the petal width is correlated with the petal length; the petal length is not correlated with the sepal width. We can confirm this in Figure 2.18.
- Look where the samples plot: grouping indicates communality between these samples.
- Look where samples plot relative to arrows: the *Iris setosa* group have very similar petal length and petal width.
- Look at the length of the arrows: the sepal length is not very important in differentiating samples.

 Visit **Notebook 06: Biplot** to visualize the blue jay and iris dataset in biplots.

Visualizing Compositional Data in a Two-Dimensional Plot

Now we turn to plotting compositional data in a two-dimensional biplot. The method is as follows:

- Perform a log-ratio transformation on the composition, or some subcomposition.
- Use the above method to create the biplot of the log ratios.

Figure 2.20 shows this for the groundwater quality dataset in the Central Valley. The clr transformation was used. What do we notice?

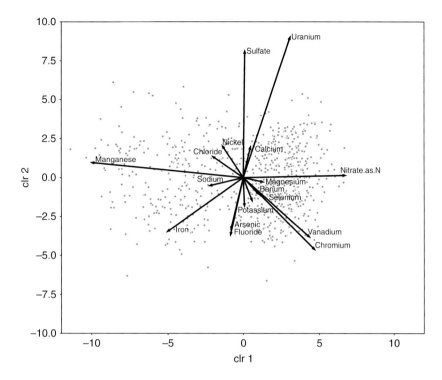

Figure 2.20 Biplot of the compositional data for the Central Valley

- We notice at least two groups in our data.
- We notice that arrows of the transition metals such as chromium and vanadium align. Manganese (Mn) is on the opposite side from chromium and vanadium.
- We notice that arsenic and chromium vary independently of each other (the arrows are orthogonal). Uranium and chromium also vary independently.

What does this mean? Well, you may not be a geochemist, but now we can approach a geochemist and ask for an explanation! Who knows, they may be pleasantly surprised by your analysis.

 Using CoDaPack, make biplot of the Central Valley. Please refer to CoDaPack_tutorial .pdf.

2.5.6 Protocol for Compositional Data Analysis

Box 2.1 presents the protocol for compositional data analysis: how to do exploratory data analysis on multivariate compositional data.

Box 2.1 The protocol for compositional data analysis

- Decide on what to do with samples that have zeros
- Ask a domain expert about interesting ratios to look at for the problem at hand
- Make log-ratio transformations and perform a basic exploratory data analysis
- Calculate the variance array and look for small and large variances
- Produce ternary diagrams for those ratios with small and large variances
- Create the biplot
- Interpret the biplot looking for alignment and orthogonality, and the length of the lines belonging to elements
- Communicate your findings with the expert, helping you in refining your analysis

Plots to make
- Exploratory data analysis of alr and clr log ratios
- Ternary diagrams and variation array
- Biplot

 Play **Video 03: Compositional Data Analysis** to learn why compositional data are special and how to do exploratory data analysis on multivariate compositional data.

2.5.7 What Have We Learned in Section 2.5?

- We have learned that the closed sum in compositions creates an artificial correlation.
- We have learned that this problem can be solved by using log ratios.
- We have learned to do an exploratory data analysis with composition, using the barycenter, total variance, and variation array.
- We have learned that compositional or multivariate data can be plotted in two dimensions by means of a biplot, where we can observe correlations between variables, or groupings that may be indicative of processes that cause these groups.

2.6 Multivariate Outlier Detection

An outlier is an anomalous value that can be of significant interest or needs to be removed from the dataset. Extremes are not outliers since they can be predicted from lesser extreme values. Outlier detection methods allow their identification, they don't explain them. The explanation is important, but how much you explain is up to you. The outlier can simply be an "error" of any kind or it can be what you are looking for! You will learn about the most common ways of performing multivariate outlier detection using the robust Mahalanobis distances, as well as being given a brief introduction to modern-day alternatives using machine learning.

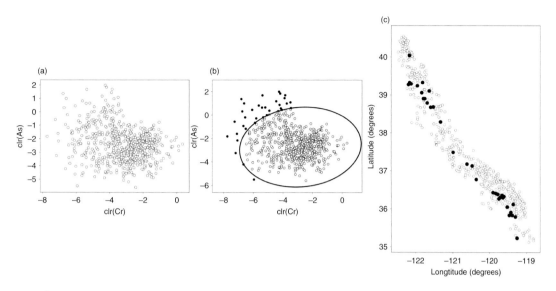

Figure 2.21 (a) Bivariate dataset of two centered log ratios of chromium (Cr) and arsenic (As) concentrations in groundwater quality data of the Central Valley; (b) robust Mahalanobis distance outliers marked in black, (c) robust Mahalanobis Distance outliers (Cr and As) marked in black at their spatial locations in the Central Valley.

Figure 2.21 shows an example of what we want to achieve. A bivariate dataset between two log ratios shows a number of sample points that appear to deviate from the bulk of samples. Multivariate outlier detection will identify these sample points. Then the location of these points can be marked on a map. Any cluster of anomalies may be indicative of an important process that is taking place, for example, an anthropogenic cause.

2.6.1 Bivariate Outlier Detection

Intuitively, an anomaly is an observation that deviates from what one may be expecting. Hence, to define outliers, we need to define what is expected. The latter is a subjective decision and therefore outlier detection is a subjective methodology. Let's consider an obvious case regarding bivariate data, in Figure 2.22. We expect a linear relationship, yet one value is clearly anomalous. In Section 2.5.5, we discussed how bivariate data often follow an elliptical scatter (this is the expectation). We will use this idea to identify the outlier. We can't identify it just by looking independently at each axis, in Figure 2.22. Indeed, a univariate outlier is not necessarily a bivariate outlier, as shown in Figure 2.23.

Mahalanobis Distance

To identify the outlier, we show what happens to the outlier after the rotation and stretching operations. All points that lie within the elliptical cloud are transformed into points that lie within a circle, except for the outlier. This suggests that we can use the concept of distance from the center of the cloud after making these transformations. The Euclidean distance measures

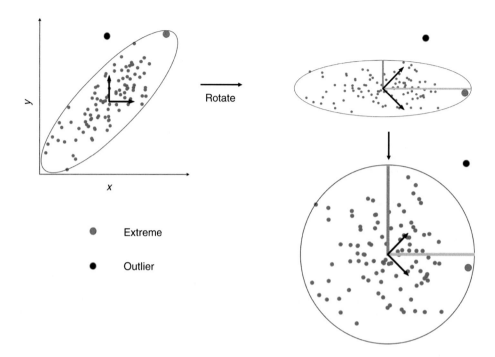

Figure 2.22 The bivariate data are rotated, then stretched. Everything outside the circle is now easily identifiable as an outlier.

Figure 2.23 Univariate outliers are not necessarily bivariate outliers. Univariate outliers are outside the rectangle, bivariate outliers are outside the ellipse.

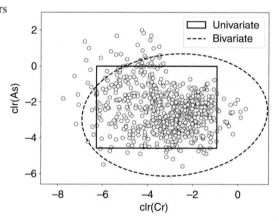

the distance between any two points in space. If we have the coordinates of two people, say $\mathbf{x}_1 = (x_1, x_2)$ and $\mathbf{x}_2 = (y_1, y_2)$, sitting in a room, they are the following distance apart:

$$d(\mathbf{x}_1, \mathbf{x}_2) = \sqrt{(x_1 - y_1)^2 + (x_2 - y_2)^2} \qquad (2.28)$$

The Euclidean distance is such that all the people sitting on a circle centered around one specific person are at equal distance. Now consider these people sitting along a circle, and now squash this circle into an ellipse. In terms of the Euclidean distance, those people that were sitting at an equal distance from the center no longer do so. However, what if we wanted to create a new distance, a form of elliptical distance, such that they still sit at the same distance from the center? This distance is termed the Mahalanobis distance (MD). To see what exactly this looks like, we can use matrices to quantify the transformation from a circle to an ellipse:

$$d_{MD}(\mathbf{x}_1, \boldsymbol{\mu}) = \sqrt{\begin{pmatrix} x_1 - \mu_1 & x_2 - \mu_2 \end{pmatrix} \begin{pmatrix} a & 0 \\ 0 & b \end{pmatrix} \begin{pmatrix} x_1 - \mu_1 \\ x_2 - \mu_2 \end{pmatrix}} \tag{2.29}$$

where $\boldsymbol{\mu}$ are the coordinates of the person at the center of the ellipse. When $a = 1$ and $b = 1$, we obtain the Euclidean distance. We can go one step further and add a rotation θ to the ellipse

$$d_{MH}(\mathbf{x}_1, \boldsymbol{\mu}) = \sqrt{\begin{pmatrix} x_1 - \mu_1 & x_2 - \mu_2 \end{pmatrix} \begin{pmatrix} cos(\theta) & -sin(\theta) \\ sin(\theta) & cos(\theta) \end{pmatrix} \begin{pmatrix} a & 0 \\ 0 & b \end{pmatrix} \begin{pmatrix} x_1 - \mu_1 \\ x_2 - \mu_2 \end{pmatrix}}$$

$$= \sqrt{\begin{pmatrix} x_1 - \mu_1 & x_2 - \mu_2 \end{pmatrix} \begin{pmatrix} c_{11} & c_{12} \\ c_{21} & c_{22} \end{pmatrix} \begin{pmatrix} x_1 - \mu_1 \\ x_2 - \mu_2 \end{pmatrix}} \tag{2.30}$$

or, in matrix notation:

$$d_{MH}(\mathbf{x}_1, \boldsymbol{\mu}) = \sqrt{(\mathbf{x}_1 - \boldsymbol{\mu})^T C (\mathbf{x}_1 - \boldsymbol{\mu})} \tag{2.31}$$

Transformations like these are also termed affine transformations. This equation applies for any dimension of the vector $\mathbf{x}$.

The ellipse we deal with is not just any ellipse, but one related to the correlation between the data, see Figure 2.21. Theory shows that the Mahalanobis transformation for bivariate data is as follows:

$$C = S^{-1}; S = \begin{pmatrix} var(X_1) & cov(X_1, X_2) \\ cov(X_1, X_2) & var(X_2) \end{pmatrix} \tag{2.32}$$

Hence, for multivariate data, the matrix is the inverse of the variance–covariance matrix S, and we get the center

$$d_{MD}(\mathbf{x}_1, \boldsymbol{\mu}) = \sqrt{(\mathbf{x}_1 - \boldsymbol{\mu})^T S^{-1} (\mathbf{x}_1 - \boldsymbol{\mu})} \tag{2.33}$$

Robust Mahalanobis Distance

The MD is a very elegant theoretical idea; in practice, however, we hit a snag. Variance–covariance matrices, just like variances and correlation coefficients, are very sensitive to outliers. This means that calculations of them are not robust, they change if we remove one or more outliers. You may think: no problem, let's remove them. Unfortunately, that is exactly what we are trying to do! We need a way out of this impasse (Americans call it a catch 22). The

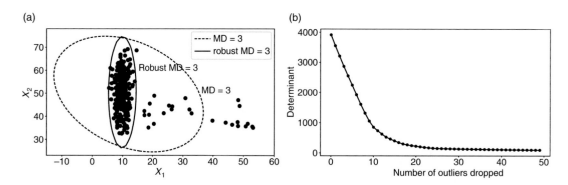

Figure 2.24 (a) Bivariate dataset that is contaminated by outliers. The correlation is zero if we remove the outliers; with the outliers, there is negative correlation. (b) Determinant of the variance–covariance as function of the number of outliers dropped.

solution lies in the determinant of the variance covariance–matrix. The determinant of a bivariate covariance matrix is as follows:

$$\det\begin{pmatrix} \text{var}(X_1) & cov(X_1,X_2) \\ cov(X_1,X_2) & \text{var}(X_2) \end{pmatrix} = \text{var}(X_1)\text{var}(X_2) - cov(X_1,X_2)^2 \qquad (2.34)$$

If no correlation exists, and all variances are 1, then we get

$$\det\begin{pmatrix} 1 & 0 \\ 0 & 1 \end{pmatrix} = 1 \qquad (2.35)$$

which is the maximum possible value; with perfect correlation, we get 0, the smallest possible value. With no correlation, bivariate data will be distributed as a circle, the maximum possible volume, while with perfect correlation we end up with a line: zero volume. Now consider the dataset in Figure 2.24. We calculate the determinant of all the data and find it to be around 4000, which is large! This is because of the outliers to the right of the main cloud of points. Let's now start dropping samples that are furthest from the center as measured by the MD. As we drop these values, the determinant becomes smaller, up to a point where the 20–30 furthest values are removed. At that point, the determinant stabilizes. The robust MD is calculated this way. The robust MD uses samples that are not outliers to estimate the variance–covariance matrix. The determinant of the variance–covariance matrix of the data aids in determining these samples.

Another interesting property of the MD has been established from theory: for data that display like ellipses, the distribution of the MD is a chi-square distribution. Hence, we can use the chi-square quantile plot to identify outliers as well (Figure 2.25).

2.6.2 Protocol for Robust Multivariate Outlier Detection

This section provides the material concepts needed to develop a full robust multivariate outlier detection method and protocol, as outlined in Box 2.2.

Box 2.2 The protocol for robust multivariate outlier detection

- If working with compositional data, make log-ratio transformations
- Use quantile-quantile plots to detect univariate outliers
- Make bivariate scatter plots of quantities of interest, including the biplot
- Inspect the determinant as a function of dropped outliers
- Make the outlier detection plot, including a high quantile (e.g., 95%) of the robust MD
- Make the chi-square quantile plot of the MD to further refine the analysis
- Decide an "outlier" vs a "non-outlier"
- Mark the outliers in their context, for example, spatial locations

Plots to make
- Quantile–quantile plot, scatter plot, clr biplot
- Determinant function of outliers removed
- Chi-square quantile plot of the robust MD
- Scatter plot indicating the robust MD
- Outliers marked on maps

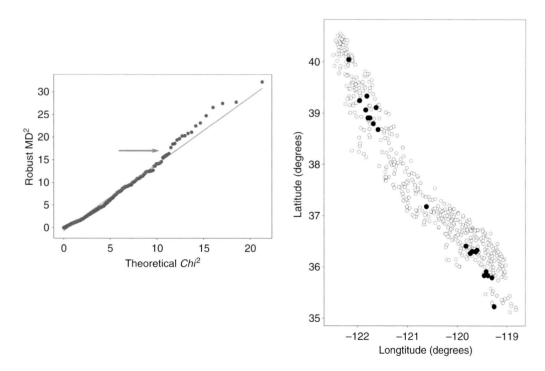

Figure 2.25 (a) Chi-square quantile plot of the robust MD. (b) Plot of samples that deviate from the straight line on a map. The black dots are outliers.

▶   Play **Video 04: Outliers** to learn about outlier detection methods for multivariate datasets.

2.6.3 Outlier Detection Using Machine Learning

Outlier detection methods identify samples in a multivariate dataset that behave differently from how we expect. With the MD, this expectation is formulated using an ellipse. Not all data behave like this; hence, there is room to either set different expectations or to take a different approach altogether. The following are a few popular options to try:

- One-class support vector machine: here the expectation is that all samples fit within some hypervolume, except that now we do not assume it to be ellipsoidal. Instead, we use an algorithm to fit such a hypervolume to the data, then any sample falling far outside this volume can be an outlier candidate.
- Local outlier factor: here we use the concept of k-nearest neighbors to determine what to expect. Using the k-nearest neighbors of each sample point, we can estimate the local density of points. For outliers, we expect them to have a substantially lower density than their neighbors.
- Isolation forest: here we randomly select a variable among all multivariate variables, then randomly select a split value between the maximum and minimum values of the selected variables. Then we divide samples based on the randomly selected value of the selected variable. We repeat all processes above until each sample has been isolated from other samples. Outliers require few steps to be isolated, because they are different from the majority of the samples.

Figure 2.26 shows three different outlier detection methods on the bivariate Central Valley dataset.

2.6.4 Application to the Central Valley Dataset: Outlier Detection

    Visit **Notebook 07: Outlier Detection** to perform outlier detection and removal on the Central Valley dataset.

2.6.5 What Have We Learned in Section 2.6?

- We have learned that univariate outliers may not be multivariate outliers and vice versa.
- We have learned that multivariate outliers may be due to chemical processes, and therefore may be of considerable interest.
- We have learned that additional interpretation is needed after multivariate outlier detection to decide what to do with them.
- We have learned about the robust MD that can detect samples that lie far from the center of the data.

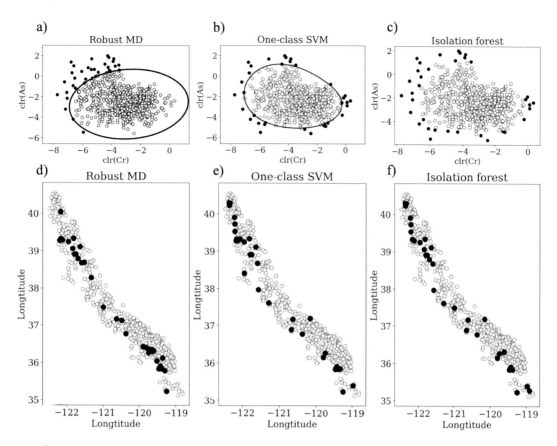

Figure 2.26 Three different bivariate outlier detection methods performed on the Central Valley dataset: (a) robust MD, (b) one-class support vector machine, and (c) isolation forest, all with bivariate outliers marked in black; (d), (e), and (f) show the outliers for (a), (b), and (c) marked in black at their spatial locations in the Central Valley.

2.7   Principal Component Analysis

Understanding multivariate data or compositional data is challenging because we are dealing with many variables at the same time. So far, insights have been obtained by making correlation tables, variation arrays, or biplots. The biplot is powerful because it plots all data in one plot, and you can get useful insights. Still, it is limited in what it can provide, so we need to take this to the next level using principal component analysis (PCA).

2.7.1   Principal Components for Bivariate Data

We should remind ourselves that correlation is what is of interest; we want to use data to make conclusions about chemical or physical processes that have taken place. At the same time, correlation indicates that there is some redundancy in our data on these processes. For example, if iron concentration is perfectly correlated with magnesium concentration, we would not need

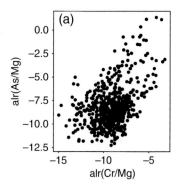 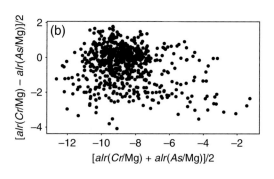

Figure 2.27 Representing data by means of linear combination, here the average sum and difference: (a) original bivariate data; (b) average sum and difference of the bivariate data.

the magnesium measurement at all; it is redundant. This suggests that correlation may help us in expressing the same dataset with much fewer variables. To get a bit more insight into why this is possible, consider Figure 2.27. The scatter plot indicates that correlation exists between these two additive log ratios. Now, let's calculate the average sum and difference between these log ratios for each sample, which results in the plot in Figure 2.27b. We can make two observations:

- the sum and the difference are not correlated
- the sum has more variance than the difference

What this suggests is that there may be simple transformations of the data that create new variables that are linearly uncorrelated, and a few variables may contain most of the variation. A more general idea that allows moving beyond equally weighted averages is as follows:

$$v_1 = a_1 x_1 + a_2 x_2$$
$$v_2 = b_1 x_1 + b_2 x_2 \tag{2.36}$$

Or, in matrix form

$$\begin{pmatrix} v_1 \\ v_2 \end{pmatrix} = \begin{pmatrix} a_1 & a_2 \\ b_1 & b_2 \end{pmatrix} \begin{pmatrix} x_1 \\ x_2 \end{pmatrix} \tag{2.37}$$

with x_1, x_2 the variables in a bivariate dataset.

Consider a hypothetical bivariate dataset as shown in Figure 2.28. This dataset shows a linear correlation and if we looked at the histogram of each variable, it would likely appear Gaussian. The data are also centered around $(0, 0)$. Our scatter plot geometrically looks like an ellipse. We recall from the biplot analysis that we can find the two axes of this ellipse. These two axes are termed principal components. The largest of these components lies along the direction where the bivariate data vary most. The other is orthogonal to it.

If we want to go from a large number of variables to a lesser amount, in a two-dimensional case this would mean going from two variables to one variable. This is called dimension

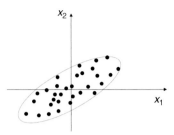
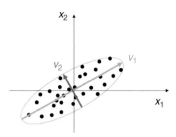

Figure 2.28 The vectors v_1 and v_2 represent two principal components.

reduction, which involves a projection, as shown in Figure 2.29. To do this, we make an orthogonal projection of the data onto the first principal component. This means that a sample $(x_{i,1}, x_{i,2})$ in the x-coordinates is represented as $(v_{i,1}, 0)$ in the v coordinates. In other words, we have taken a sample with two values and made it a sample with only one value: $v_{i,1}$. Of course, we have changed our sample, we made an approximation, or we are making an error that way. Hence, the question is how much change was made or, in other words, how much information did we lose in doing so?

To understand what error we are making, let's look at the length of the vector of the projection, namely the lengths of ε in Figure 2.30. The longer ε, the greater the error. So, if we sum up the all the lengths ε for all samples, we have a measure of the error, which is termed the "variance of the error." How do we calculate these lengths? Very simple: we project onto the second principal component, the light-grey (orange) axis. Now we go from $(x_{i,1}, x_{i,2}) \rightarrow (0, v_{i,2})$, where $v_{i,2}$ is the amount of error we make by projecting onto the dark-grey (green) axis.

Imagine now doing the opposite: we project onto the second principal component and would like to know how much information we are losing in doing so. Obviously, we are losing much more information than projecting onto the first principal component. For one sample this error is now equal to $v_{i,1}$. Hence, if we add that up for all samples, the total amount of information lost is simply the variance observed on the first principal component vector.

Since both variances are calculated along the orthogonal axis, we can calculate a total variance:

$$\text{Total variance} = var(v_1) + var(v_2) \qquad (2.38)$$

or we can represent this by relative contributions:

$$\text{variance contribution along } v_1 = \frac{var(v_1)}{var(v_1) + var(v_2)}$$

$$\text{variance contribution along } v_2 = \frac{var(v_2)}{var(v_1) + var(v_2)} \qquad (2.39)$$

2.7.2 Principal Components for Multivariate Data

Now we are ready to move to multivariate data. We do not cover any derivation; instead, we focus on what any PCA code outputs, what that means, and how to interpret it. We'll consider a four-variate subcomposition of the Central Valley dataset, as shown in Figure 2.31.

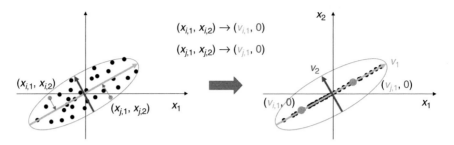

Figure 2.29 Projecting samples onto the first principal component vector, the dark-grey (green) axis.

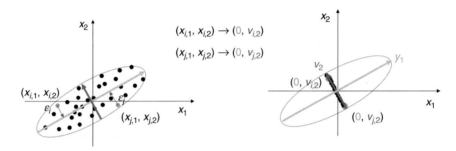

Figure 2.30 Projecting samples on the second principal component vector, the light-grey (orange) axis.

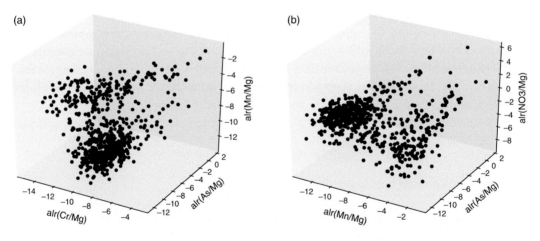

Figure 2.31 Two scatter plots of a four-variate compositional dataset with additive log ratios. Each scatter plot has three variates.

Information Generated by PCA

Recall from the bivariate case that principal component vectors are weighted averages of the original data. With four variables, this becomes:

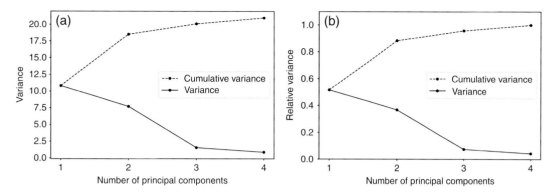

Figure 2.32 Scree plots ℓ_{ij} of the four-variate compositional dataset in Figure 2.31. Dashed lines are cumulative variance, while solid lines are variance contribution for each principal component i: (a) scree plot, absolute variance; (b) scree plot, relative variance.

$$v_1 = \ell_{11}x_1 + \ell_{12}x_2 + \ell_{13}x_3 + \ell_{14}x_4$$
$$v_2 = \ell_{21}x_1 + \ell_{22}x_2 + \ell_{23}x_3 + \ell_{24}x_4$$
$$v_3 = \ell_{31}x_1 + \ell_{32}x_2 + \ell_{33}x_3 + \ell_{34}x_4$$
$$v_4 = \ell_{41}x_1 + \ell_{42}x_2 + \ell_{43}x_3 + \ell_{44}x_4 \tag{2.40}$$

In PCA, we called these weights "loadings"; hence, we use ℓ, where ℓ_{ij} is the loading for principal component i on variable x_j. Any PCA code will output these values.

As with the bivariate case, we can calculate variances of the v_i. Now, we have four such variances: a variance of 10.82 for v_1, 7.71 for v_2, 1.57 for v_3, and 0.89 for v_4; and a variance of 3.14 for x_1, 4.58 for x_2, 8.41 for x_3, and 4.83 for x_4.

Variances of principal components are usually organized in a scree plot; see Figure 2.32. The scree plot ranks the variances of each principal component from large to small (solid line) as well as the cumulative variance (dashed line). The cumulative function helps determine what the proportion of variance is as function of the number of principal components. This will help in determining the approximation made by any dimension reduction. For example, if only three of the four dimensions are retained, what is the variance retained in three dimensions relative to the total variance? Similarly: what is the variance removed when going from four to two dimesions?

Because principal components no longer have any linear correlation, we can add the variances of the v_i to obtain a total variance:

$$var_{tot} = var(v_1) + var(v_2) + var(v_3) + var(v_4) \tag{2.41}$$

This value can be useful to compare two datasets. It answers the question of which multivariate dataset has more variance.

Next, we plot the loadings to investigate the meaning of each principal component, see Figure 2.33. For example, in the principal component vector 1 (PC1), we observe that $alr(Cr/Mg)$ and $alr(NO_3/Mg)$ are given small weights, while the other are given positive weights, so this PC1 is an average of As and Mn. PC2 is also an average of Cr, As and nitrate (NO_3),

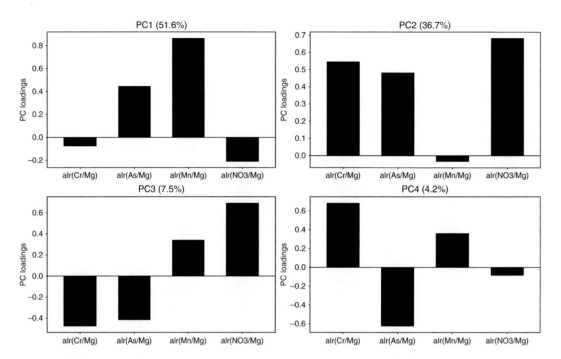

Figure 2.33 Loadings ℓ_{ij} corresponding to each principal component i. The percentage shows the relative explained variance for each PC.

while PC3 averages Cr with As, Mn with NO_3, then calculate the difference between the two averages.

Next, we make score plots with v_i. Principal component scores are projections onto a principal component axis; they are coordinates of one sample on one principal component axis. For example, if we have a sample $(x_1, x_2, x_3, x_4) = (-2, -5, -7, +1)$ of log ratios, then the score v_1 of that log ratio on the first principal component is

$$v_1 = \ell_{11}(-2) + \ell_{12}(-5) + \ell_{13}(-7) + \ell_{14}(+1) \tag{2.42}$$

We can calculate this for all principal components. In a score plot, we plot these for two selected principal components, for example PC1 vs PC2, PC1 vs PC3, etc. Typically, we focus on scores on the lower principal components, which represent most of the variation in the multivariate data. Figure 2.34 shows a plot of the first vs the second principal component scores of our data. In this scatter plot, we are looking at about 90% of all the variation in the data.

In score plots, one can also color or shade each point with some value of interest. Recall that each point in this plot is a sample value. The score plot may reveal interesting features in the data. For example, in Figure 2.34, we observe how two clusters are emerging. These clusters may then warrant further geochemical interpretation and analysis: why are there clusters? What do they represent? Additionally, we notice in each cluster that there is a variation in the shading. In the left cluster, it is constant; in the right cluster, however, the shading has a trend. Again, we should go back to our geochemical data and case, and aim to find interpretations for this.

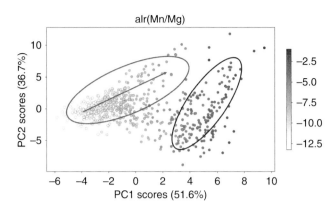

Figure 2.34 Score plot between PC1 and PC2; the shaded scale is the value of alr(Mn/Mg).

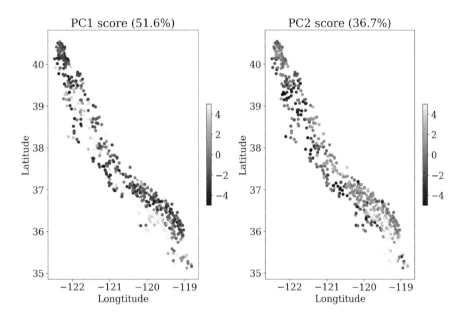

Figure 2.35 Mapping principal component scores onto the sample locations. A black and white version of this figure will appear in some formats. For the color version, refer to the plate section.

Finally, we should put the principal component scores into the map of where the data are coming from; see Figure 2.35. In doing so, we now observe "spatial correlation," meaning that high values (or low values) seem to be at locations near each other. Again, this is likely to have an interesting geochemical interpretation.

Standardizing Data

Here we discuss the importance of standardizing the dataset before the actual PCA. Some codes may have this as default, others require you to specify this choice. In PCA, we are interested in correlation. We exploit this correlation to find important directions (vectors) in the dataset. Lower principal components represent directions with significant variations (variance). However, variance is function of the unit used. Some variables may be in kilograms, others in millimeters. Recall also that linear correlation coefficients (Eq. (2.2)) are standardized variables. To remove the dependency on the units and quantify the variance in the multivariate dataset based on correlations, we need to standardize the dataset.

Dimension Reduction Using PCA

PCA provides a bijective transformation between the data and the principal component scores. This means that there is a unique way to go from the x to the v and vice versa. Any PCA will therefore also output values of τ, where we can calculate τ from Eq. (2.40):

$$x_1 = \tau_{11}v_1 + \tau_{12}v_2 + \tau_{13}v_3 + \tau_{14}v_4$$

$$x_2 = \tau_{21}v_1 + \tau_{22}v_2 + \tau_{23}v_3 + \tau_{24}v_4$$

$$x_3 = \tau_{31}v_1 + \tau_{32}v_2 + \tau_{33}v_3 + \tau_{34}v_4$$

$$x_4 = \tau_{41}v_1 + \tau_{42}v_2 + \tau_{43}v_3 + \tau_{44}v_4 \tag{2.43}$$

What is dimension reduction? It means, for example, a reduction from four dimensions to two dimensions, where we approximate the data as follows:

$$\widetilde{x}_1 = \tau_{11}v_1 + \tau_{12}v_2$$

$$\widetilde{x}_2 = \tau_{21}v_1 + \tau_{22}v_2$$

$$\widetilde{x}_3 = \tau_{31}v_1 + \tau_{32}v_2$$

$$\widetilde{x}_4 = \tau_{41}v_1 + \tau_{42}v_2 \tag{2.44}$$

In other words, we cut down the sum. The dimension reduction results in an error, and it is important to evaluate what this error is for each sample in the dataset so we can compare a sample $\mathbf{x}$ with a sample $\widetilde{\mathbf{x}}$. For each sample, the absolute error is:

$$\varepsilon_{abs} = \|\mathbf{x} - \widetilde{\mathbf{x}}\| \tag{2.45}$$

Or we can calculate the relative error:

$$\varepsilon_{rel} = \frac{\|\mathbf{x} - \widetilde{\mathbf{x}}\|}{\|\mathbf{x}\|} \tag{2.46}$$

We can analyze this relative error by shading the score plot with the value of the relative error (Figure 2.36). Samples with high PC1 and PC2 scores have large relative errors if we reduce the number of dimensions to two.

2.7.3 Limitations of PCA

PCA relies on linear correlation in datasets. Hence, when the data contain non-linear correlations, the results may become less easy to interpret. Apart from a few pathological cases

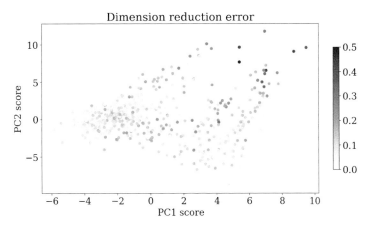

Figure 2.36 Score plot showing the dimension reduction, with the value of the relative errors shaded according to the scale shown. A black and white version of this figure will appear in some formats. For the color version, refer to the plate section.

however, PCA is still useful for multivariate data that show non-linear correlation: it will still create an orthogonal axis system and allows for dimension reduction; what becomes less effective is the interpretation of the principal component vectors.

2.7.4 Protocol for PCA

Box 2.3 presents the protocol for PCA, including the preprocessing steps and visualizations on PCA loading and score plots.

Box 2.3 The protocol for PCA

- If working with compositional data, make log-ratio transformations (preferably ilr)
- Standardize data to make sure variables are on the same scale
- Run the PCA code
- Make a scree plot to understand the variance contributions of each principal component vector
- Make loading plots to attribute meaning to principal components
- Make score plots, and shade or color them with quantities of interest
- Map the scores back onto the location in space where the sample was collected
- When performing dimension reduction, map the lower-dimensional score back to the original variables and assess the error

Plots to make

- Scree plot
- Loading plots
- Score plots
- For spatial data: maps with principal component scores
- For dimension reduction: error made for each sample

Play **Video 05: Principal Component Analysis and Factor Analysis** to learn the PCA protocol.

2.7.5   Application to the Central Valley Dataset: PCA

 Visit **Notebook 08: Principal Component Analysis** to perform PCA on the Central Valley dataset.

2.7.6   What Have We Learned in Section 2.7?

- We have learned that PCA is an analysis that shows which directions in the data have large variations and which have small variations.
- We have learned to interpret the meaning of these directions through loading plots. In such loading plots, we show how each principal component direction is a weighting of the original variables.
- We have learned to interpret the scree plot, which shows the ranking of the principal component directions in terms of how much variance they explain.
- We have learned to interpret score plots, in which we can shade or color samples with a property of interest.

2.8   Cluster Analysis

So far, we have focused on using variations in different variables to analyze multivariate data. There is an alternative way to study relationships in multivariate datasets by means of distances, which means how far apart different samples are. Now, instead of looking at similarities between variables, we are looking at similarities between samples. If a set of samples are very similar, that may be indicative of a process. You can also view this in a mathematical sense. For example, in a composition with D components, of which we have N samples, there are two dimensions: D and N. With variables, we operate in the D dimension; when working with sample similarities, we operate in the N dimension.

Cluster analysis uses a measure of similarity between samples to divide these samples into distinct (mutually exclusive) groups. We are interested here in unsupervised clustering methods. This means that in the dataset the samples do not have any kind of indicative labels attached to them that tell us what group they may belong to. In unsupervised clustering, we need to discover those labels from the data, including the number of groups.

A key component in clustering is the notion of similarity, which mathematically is formulated as a distance, or dissimilarity. A very popular clustering method that uses distance is k-means clustering.

2.8.1 *k*-Means Clustering

Among many available clustering algorithms, *k*-means clustering is probably one of the most widely used due to its simplicity and efficiency. The *k* here refers to the number of clusters. In *k*-means clustering, you will need to specify that number; however, in the next section, we will discuss statistical ways to discover this number from the data.

We illustrate clustering on the same dataset used to illustrate PCA, shown in Figure 2.31,. Here, we have a composition so, as usual, we need to make a log-ratio transformation. A common distance used in defining the similarity between any two samples $\mathbf{x} = (x_1, \ x_2, \ ..., \ x_D)$ and $\mathbf{y} = (y_1, \ y_2, \ ..., \ y_D)$ is the Euclidean distance:

$$d(\mathbf{x}, \mathbf{y}) = \sqrt{(x_1 - y_1)^2 + (x_2 - y_2)^2 + \ldots + (x_D - y_D)^2} \tag{2.47}$$

However, we cannot calculate the Euclidean distance directly from the compositional data, for the same reason that we cannot apply it to any other statistical summary such as the mean, variance, or correlation coefficient. A log-ratio transformation is needed. Here, we need to use the ilr transformation, briefly mentioned in Section 2.5.4. Although we did not cover the mathematical details of this log ratio, we can still understand its usefulness through the notion of "isometry." Isometric transformations in mathematics have the property of preserving shapes. This property also entails that distance properties are preserved. For example, if two samples are equidistant from a third sample, after transformation that property is preserved. Since clustering requires stating distances, then this property of the ilr is exactly what we need to perform cluster analysis of compositions!

After the ilr transformation, we can use the Euclidean distance in the new space we created. Only two inputs are required for *k*-means clustering:

- the distance: we use the Euclidean distance after ilr transformation
- the number of clusters: we use two clusters for the reason explained in the next section

k-Means clustering then uses the distance metric and the number of clusters specified by users. First, the method randomly assigns each data point to one of the *k* clusters. Then, it calculates the mean of the samples in each cluster. For each data point, the method calculates the distance between itself and the mean of each cluster, and reassigns the point to the cluster with the smallest distance. Next, the algorithm recalculates the mean of the samples in each updated cluster, and repeats the distance calculation and reassignment for each data point. This process continues until there is no change in the assignment of data points to clusters.

The outcome of *k*-means clustering is that each sample now has a label attached: the cluster the sample belongs to. As shown in Figure 2.37 we can attach these labels to the spatial locations where samples were taken and approach a geochemist to ask if the observed distribution of these labels conveys any information about process.

Now that we have the result, we will dive a bit deeper into analyzing what we did. This analysis will focus on (1) analyzing the distance we used and (2) determining a suitable number of clusters.

Figure 2.37 Labels that are returned by k-means clustering of the data in Figure 2.31. The number of clusters is fixed at two: the dark (purple) circles and the light (yellow) circles indicate samples from two different clusters. The distance is the Euclidian distance after the ilr log-ratio transformation.

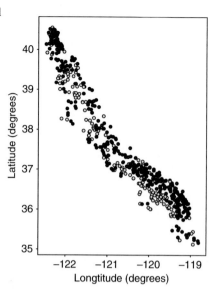

2.8.2   How Many Clusters Are in My Datasets?

In the k-means procedure, the number of clusters k needs to be specified prior to clustering. What is a good number here? Let's look at two extremes: putting all samples into one cluster; and making as many clusters as samples. Clearly neither are good choices, so perhaps there is some good middle way. Ideally, we'd like clusters to be compact and distinct, meaning that samples within a cluster are closer to the middle of the cluster than to the middle of other clusters. This is not guaranteed! We need to make sure that this criterion is followed as much as possible. We can start with calculating two important properties:

- Compact: $a(i)$: the average distance of a sample i to all other samples that are in the same cluster. This measures "how well" a sample belongs to that cluster. A small $a(i)$ means close to the center.
- Distinct: $b(i)$: the minimum of the average distance of a sample i to samples in clusters of which that sample is not a member. The cluster with the smallest average distance is considered to be the neighbor cluster. The larger $b(i)$, the further the neighbor cluster is, and hence the more distinct the clusters are.

We can combine the two criteria into one metric for sample i as

$$s(i) = \frac{b(i) - a(i)}{\max(a(i), b(i))} \tag{2.48}$$

The standardization means that $-1 \le s(i) \le 1$. The silhouette index is the average of the $s(i)$ for each sample, and is a function of the number of clusters k:

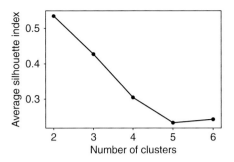

Figure 2.38 Silhouette index identifying two as the optimal number of clusters.

$$S(k) = \frac{1}{N} \sum_{i=1}^{N} s_k(i) \tag{2.49}$$

We are looking for this index to be large. Figure 2.38 shows the silhouette index for our illustration dataset, and indicates that two clusters are a good choice.

2.8.3 *k*-Medoid Clustering

We have learned that the median is a 50% quantile, it is the "middle data value." Also, the median, unlike the mean, is an actual value in the sample. In *k*-means clustering, we calculate the mean of the samples in a cluster. Hence, that mean is not a sample in our dataset. To extend the concept of median to more than one dimension, we introduce the concept of medoid: the "middle sample." What is the "middle"? To be able to define this for dimensions higher than one, we need to use distances. The middle then is that sample in our dataset that is, on average, the closest to all other samples. Therefore, in *k*-medoid clustering, the output is not the mean of each cluster but the middle, or medoid, of each cluster, which is a sample of our dataset. You can also regard the medoid as a representative sample for that cluster.

2.8.4 Evaluating *k*-Means Clustering and Displaying Results

k-Means clustering provides us with flexibility, which is great: we can choose whatever distance we want based on the type of multivariate data. Such a luxury, however, like many luxuries in life, comes down to what the "best" choice is. We would prefer distances that provide compact and distinct clusters. But what if we have several candidate distances, how would we evaluate them? Next, we provide a thorough and foundational method for studying these distances, namely: multidimensional scaling.

Multidimensional Scaling

We return to the concept that a multivariate dataset can be viewed two ways: the correlation between variables and the distance between samples. In mathematics, we call this a duality. In other words, calculating the Euclidean distance between samples is the dual to calculating the variance–covariance matrix of random variables of those samples. The term dual here means that both express the same information about relationships in multivariate data.

We have learned that PCA relies on the correlation. Hence, an intriguing question is: if correlation is the dual of Euclidean distance, what then is the dual of PCA? The answer is: multidimensional scaling (MDS). Let's take an example dataset. Figure 2.39, shows the data projected in two dimensions (the score plot), one with MDS using the Euclidean distance of ilr coordinates, the other with PCA, based on the variance–covariance matrix of ilr coordinates. Each score plot conveys the following:

- In the PCA score plot, the axes are the two directions on the dataset that best explain the variance. Of all orthogonal directions, these directions contain the most variation.
- The MDS score plot is such that the distance between the points in two dimensions best approximates the distance between the actual data (which are in four dimensions). This plot best preserves the distance (Figure 2.40).

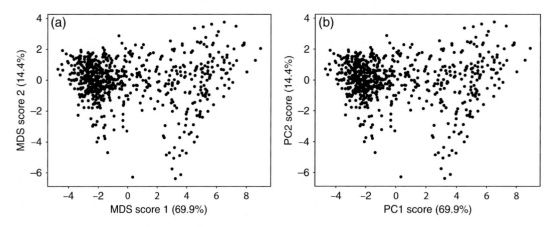

Figure 2.39 Score plots produced by (a) MDS and (b) PCA.

Figure 2.40 Comparison of actual Euclidean distance (Euclidean distances are calculated with four dimensions) and samples with a two-dimensionalal Euclidean MDS distance between the samples of the four-variate compositional dataset after ilr transformation in Figure 2.31.

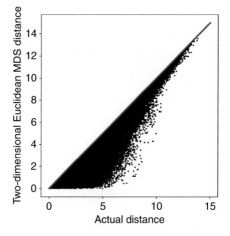

Both plots in Figure 2.39. are exactly the same. The duality works out. However, note that in MDS you can use any distance, you are not limited to the Euclidean distance. In PCA, unfortunately, we are stuck with the variance–covariance matrix.

Both PCA and MDS solve by singular value decomposition. We can also make scree plots for both cases, which are shown in Figure 2.41. These two scree plots are the same.

MDS as a Way of Visualizing Clustering Results

We have now established a link between clustering and MDS. Clustering uses a distance as measure of similarity between samples, while MDS helps to display the data in lower dimensions using that same distance. It therefore makes sense to also plot the cluster labels into the score plot of MDS, as shown in Figure 2.42.

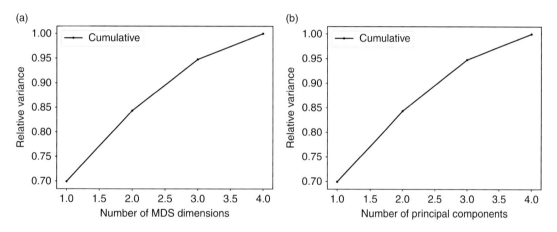

Figure 2.41 Scree plots from eigenvalue decompositions for both (a) MDS and (b) PCA.

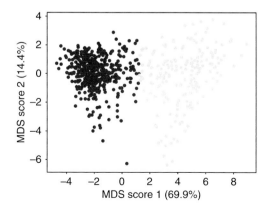

Figure 2.42 Results of k-means clustering in an MDS plot that uses the same distance as in clustering.

Box 2.4 The protocol for clustering analysis

- If working with compositional data, make log-ratio transformations (preferably ilr)
- Choose a distance that defines the difference between samples
- Perform MDS, make the scree plot
- Use the scree plot to analyze the quality of your distance
- Plot the sample data in two and three dimensions using the scores provided by MDS
- In the MDS score plots, shade or color the points with a quantity of interest
- Create a silhouette plot and decide on the number of clusters
- Run k-means clustering, plot the cluster labels in the MDS score plot
- Plot the cluster labels in the context of the data, for example, the spatial location of the data

Plots to make
- clr biplot
- MDS: scree plot and score plot
- Score plots with dots shaded or colored by the property of interest
- Silhouette plot
- MDS score plot with cluster labels

2.8.5 Protocol for Cluster Analysis

One important note in cluster analysis is not blindly trusting the distance you use, particularly when your samples have lots of variables. Therefore, we recommend performing MDS, to look at the scree plot, or to check the error in distance on increasing dimensions, as shown in Figure 2.40. A distance that still performs well in the lower dimensions after MDS is a good distance.

Box 2.4 provides the protocol for cluster analysis, including MDS, k-means methods, and how to decide the number of clusters.

2.8.6 Application to the Central Valley Dataset: Cluster Analysis

 Visit **Notebook 09: Clustering** to perform cluster analysis on the Central Valley dataset.

2.8.7 What Have We Learned In Section 2.8?

- We have learned about two ways to represent relationships in multivariate data, through correlation and through distance.

- We have learned that distances between samples allow for clustering, hence the discovery of grouping that may have real causes.
- We have learned that we can use distances between samples to make score plots using MDS.

2.9 Factor Analysis

We have mentioned several times that the aim of data analysis is to discover "reasons": what process took place? What created variations in the data? In factor analysis (FA), we aim to discover what these reasons are; we will call them factors. Data science alone does not determine physical or chemical processes. Hence, any statistical FA will need to be accompanied with an expert interpretation. We can ask domain experts to help us.

2.9.1 What Is the Purpose of FA?

Let's start with a simple example in two dimensions; see Figure 2.43. In FA, we need to first hypothesize how many different factors influence the data. Imagine we hypothesize there is one single factor f_1. We turn this idea into the following equations:

$$x_1 = a_1 f_1 + \varepsilon_1$$
$$x_2 = a_2 f_2 + \varepsilon_2$$

(2.50)

These equations state that the variables x_1, x_2 can be expressed by one single variable f_1 plus some random noise ε. The random noise could be variation in the data that we are not interested in. What FA finds is:

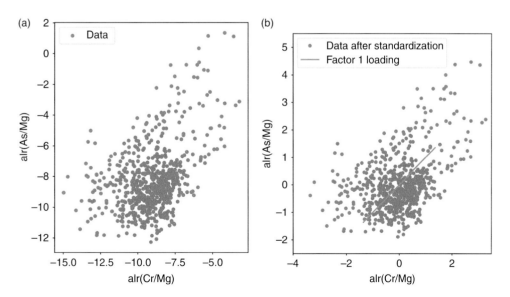

Figure 2.43 Example of FA: (a) bivariate dataset; (b) bivariate data after standardization and factor 1 loading.

- What are the "loadings": a_1, a_2?
- How much variation is expressed by f_1 versus the random noise $\varepsilon_1, \varepsilon_2$?

If we run a standard FA code, we get the loadings:

$$
\begin{aligned}
x_1 &= 0.647\, f_1 + \varepsilon_1 \\
x_2 &= 0.647\, f_1 + \varepsilon_2
\end{aligned}
\tag{2.51}
$$

and it outputs that f_1 exhibits 42% of variation, while ε_1 and ε_2 have 58% of variation. We notice that factor 1 represents a positive weighted sum of the variables. We would now approach a geochemist and discuss the meaning of this.

Let's move to a four-dimensional example, namely the iris dataset in Section 2.5.5. Assume we spoke to a botanist who conjectures that two major factors influence the variation observed in iris flowers: climate and soil chemistry. Hence, we run a two-factor FA:

$$
\begin{aligned}
x_1 &= \ell_{11}\, f_1 + \ell_{12}\, f_2 + \varepsilon_1 \\
x_2 &= \ell_{21}\, f_1 + \ell_{22}\, f_2 + \varepsilon_2 \\
x_3 &= \ell_{31}\, f_1 + \ell_{32}\, f_2 + \varepsilon_3 \\
x_4 &= \ell_{41}\, f_1 + \ell_{42}\, f_2 + \varepsilon_4
\end{aligned}
\tag{2.52}
$$

The output of a FA code are the loadings and the variances. In Figure 2.44, we observe that the first factor has high loadings on three variables, while the second has a high loading on the remaining variable.

Let's turn to a more in-depth analysis of these loadings. First, in FA we can also calculate the errors we are making, as we did in PCA by projecting each sample onto the factor vector and calculating the projected scores and errors for each variable. Figure 2.45 shows the results. Overall, the errors are quite random, except for a few combinations such as variable 3 (petal width) and variable 4 (petal length).

In FA, the loadings ℓ_{ij} are given the following meaning:

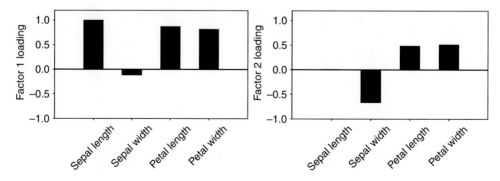

Figure 2.44 Loadings of a two-factor model for the iris dataset.

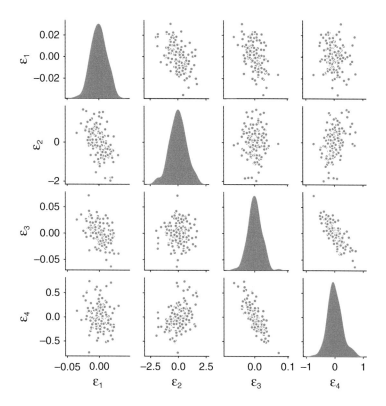

Figure 2.45 Analysis of the random error in the iris dataset with two factors.

- The magnitude of the loading on factor i for variable j: this tells us that factor (cause) i is causing a variation in variable j. If the loading is zero, that factor does not cause a variation in variable x_j,
- The square of the loadings: this represents the relative variance explained by a particular factor, and hence the cause.

In terms of squares, we can look at them row-wise and column-wise. First consider the loadings column-wise, and make the sum of squares:

$$\left(\ell_{11}^2 + \ell_{21}^2 + \ell_{31}^2 + \ell_{41}^2\right) \tag{2.53}$$

This sum is the relative variance of all four variables explained by factor 1. Similarly, we can work row-wise and calculate:

$$\left(\ell_{11}^2 + \ell_{12}^2\right) \tag{2.54}$$

This sum of squares is the relative variance of the first variable explained by the two-factor model. This relative variance explained by factors is also termed "communality," which refers to what is common among all variables, namely the factors, versus what is considered random.

Table 2.6 A typical output of a factor model

Variable x_i	Observed variance (standardized)	Loadings on factor 1: ℓ_{i1}	Loadings on factor 2: ℓ_{i2}	Communality: $\ell_{i1}^2 + \ell_{i2}^2$
Sepal length	1	0.997	0.006	0.995
Sepal width	1	−0.115	−0.665	0.455
Petal length	1	0.871	0.486	0.995
Petal width	1	0.818	0.514	0.932
Overall	4	2.436	0.942	3.378

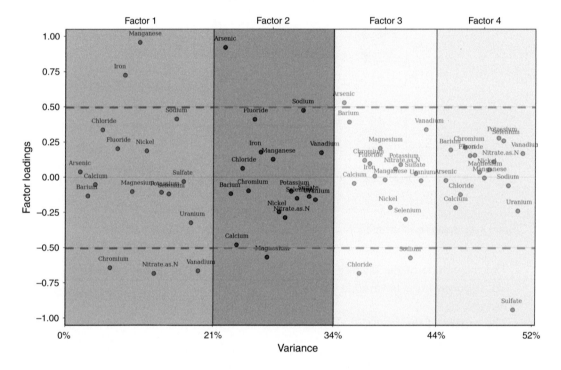

Figure 2.46 Combination of factor loadings with variances, for each factor.

Table 2.6 summarizes all the variances and communalities for easy comparison. Column-wise, we list the variables with the variances standardized to "1." In the rows we list the loadings and the communalities. Because of the standardization, these communalities are proportional to the variance. Here, we notice that all variables in the dataset, except the sepal width, are explained very well by the factor model. We can go back to the botanist and tell them that two factors explain variations in three variables very well. They may then link these to actual causes, such as climate and soil chemistry.

In cases with many variables, such as in the Central Valley cases, loadings are usually presented as shown in Figure 2.46, where two plots are combined, the scree plot and the loadings plot. Each factor is represented by a shaded box. The width of this box is equal to the relative variance explained by that factor. The *y*-axis shows the loadings. In addition, dashed lines indicate when loadings become high in absolute value.

2.9.2 FA Versus PCA

At this point you may be wondering what the difference is between FA and PCA. After all, it seems all we did was to state there is one factor and then calculate loadings and variances. At first glance, this appears similar to doing PCA and keeping only one principal component. While there is a difference in the computation, the main difference lies in the purpose.

- The purpose of PCA is to explain *total variation* in a dataset of *N* variables with *N* principal components. We may also use PCA for dimension reduction where we aim to keep the maximum variance, or minimize the loss of information.
- The purpose of FA is to explain *correlation* in a dataset of *N* variables by hypothesizing that a fixed number of *K* factors exist that explain this correlation. We are not interested in dimension reduction as an outcome of the analysis; hence we do not care (too much) about any loss of information.

If we run a standard PCA and FA code, then indeed, we get different results, as shown in Figures 2.47 and 2.48. Let's first look again at PCA:

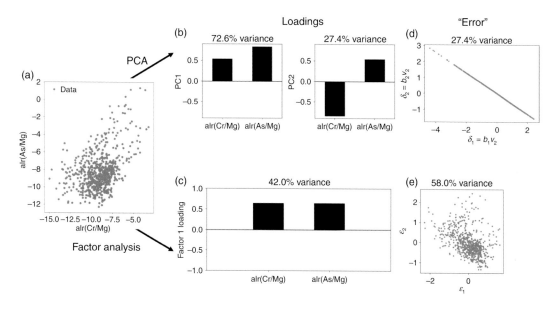

Figure 2.47 Comparison between PCA and FA: (a) bivariate data, (b) loading plots for PC1 and PC2, (c) loading plot for factor 1, (d) errors or information loss if we only keep PC1, (e) errors if we only keep factor 1.

Figure 2.48 The first factor is not equal to the first principal component (PC1), here shown after standardization of variables. Principal components PC1 and PC2 are orthogongal in the unstandardized space.

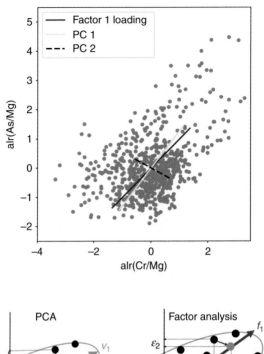

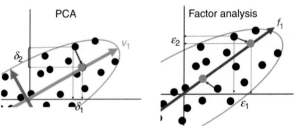

Figure 2.49 (a) In PCA, the projections are orthogonal, the δ values are correlated (b) In FA, the ε values are not correlated, hence the projection is not orthogonal.

$$x_1 = a_1 v_1 + b_1 v_2$$
$$x_2 = a_2 v_1 + b_2 v_2 \tag{2.55}$$

Dimension reduction means keeping fewer components; for example, if we keep only one component then we would make an "error," which means we lose information:

$$x_1 = a_1 v_1 + \delta_1$$
$$x_2 = a_2 v_1 + \delta_2 \tag{2.56}$$

This looks very much like a factor model, but it is not. Figure 2.47 shows that δ_1 and δ_2 are perfectly correlated. The reason for this is explained in Figure 2.49. In PCA, we make orthogonal projections on the first principal component, because this was shown to preserve maximum variation when going from two dimensions to one dimension. In the FA model, the remaining component is simply random noise. Hence, the projection is not orthogonal. Consequently, the first factor has different loading compared to the first principal component; see Figure 2.48.

2.9.3   Rotation of the Factor Loadings

An important disadvantage of the factor model is that many combinations of loading exist that explain correlation in the same data equally well. Take, for example, the following two factor models where we have three variables and speculate two factors:

$$
\begin{aligned}
&\text{Model 1}\\
&x_1 = 0.5f_1 + 0.5f_2 + \varepsilon_1\\
&x_2 = 0.3f_1 + 0.3f_2 + \varepsilon_2\\
&x_3 = 0.5f_1 - 0.5f_2 + \varepsilon_3
\end{aligned}
\tag{2.57}
$$

$$
\begin{aligned}
&\text{Model 2}\\
&x_1 = \sqrt{2}/2\,f_1 + 0f_2 + \varepsilon_1\\
&x_2 = 0.3\sqrt{2}\,f_1 + 0f_2 + \varepsilon_2\\
&x_3 = 0f_1 - \sqrt{2}/2\,f_2 + \varepsilon_3
\end{aligned}
\tag{2.58}
$$

We can calculate the variances of the variables, predicted by both models as follows:

$$
var(x_i) = \ell_{i1}^2 var(f_1) + \ell_{i2}^2 var(f_2) + \sigma_i^2
\tag{2.59}
$$

where σ_i^2 is the variance of ε_i.

In FA, we impose the following assumption: namely, $var(f_i) = 1$, or

$$
var(x_i) = \ell_{i1}^2 + \ell_{i2}^2 + \sigma_i^2
\tag{2.60}
$$

If you plug in the above numbers for the loadings, you will find that these variances are exactly the same. Nevertheless, there is a difference between factor model 1 and 2: factor model 2 is much easier to interpret because it has more zero loadings. One way to see the relationships between the loadings is to plot them in two dimensions; see Figure 2.50. We can now observe the loadings of model 2 can be obtained from model 1 by rotation. For that

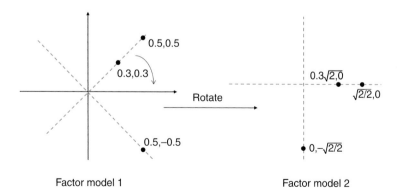

Figure 2.50 Rotation of the factor loadings creates an factor model that is easier to interpret, with more loadings equal to zero.

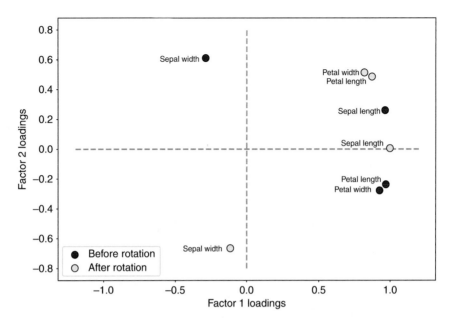

Figure 2.51 Varimax rotation of FA loadings. The varimax rotation is to maximize the total variance of the factor loadings of each factor in order to get the maximum interpretability.

reason, many FA codes include rotation options. Unfortunately (again!) you can make many types of rotations in high dimensions, which leaves still some level of subjectivity in determining the loadings.

Figure 2.51 shows how rotations work for the iris dataset. First, the FA code computes a set of loadings, then it rotates the loadings to get maximum interpretability.

2.9.4 Protocol for FA

As with PCA, it is important to follow a protocol. However, unlike PCA, in FA the major subjectivity lies in the choice of number factors, and the rotations, and here there is a possibility to cheat. People may cheat if they want to confirm a hypothesis that they have about factors, so they tune the FA in their advantage, for example by trying out different choices of factors until it aligns with their expectations! Obviously, this is poor scientific practice. If your data analysis does not confirm your expectations, then something potentially more interesting is happening than you could have imagined. The search for that explanation is where science progresses.

Box 2.5 provides the protocol for factor analysis, including deciding how many factors to use and, especially, communication with domain experts.

Box 2.5 The protocol for FA

- Communicate with the expert about possible causes that created the variations in the data
- If working with compositional data, make log-ratio transformations (preferably ilr)
- Make sure to unmask outliers and explain them with help of the expert
- Perform a computational data analysis, PCA, clustering analysis to get more insight into the dataset
- Perform FA with the number of factors hypothesized and make loading plots
- Analyze the performance of the FA: calculate communalities, assess if the remaining error is random
- Confer with the expert on the loading plots
- Project data onto factor scores, make score plots
- Plot factor scores in space on a map, confer with an expert on the meaning

Plots to make
- Computational data analysis, PCA related plots
- Loading plots
- Factor score plots
- Spatial map of factor scores

 Visit **Notebook 10: Factor Analysis** to perform FA on the iris dataset.

2.9.5 Application to the Central Valley Dataset: FA

 Visit **Notebook 11: Factor Analysis on Central Valley data** to perform FA on the central valley dataset.

Two key figures for the Central Valley dataset are Figure 2.46 and 2.52. The former shows loadings and variance, the latter shows the factor scores of each sample plotted in its geographic context. The expert on this dataset used these plots to make the following interpretation:

- Factor 1: This represents the groundwater redox conditions: the positive scores and loadings correspond with low-oxygen conditions. Manganese and iron are more mobile in their reduced states (as Mn(II) and Fe(II)) – reductive dissolution of iron and manganese solid phases typically occurs when oxygen is depleted (manganese and iron become favorable electron acceptors). Since we don't have dissolved oxygen in this composition, nitrate is the

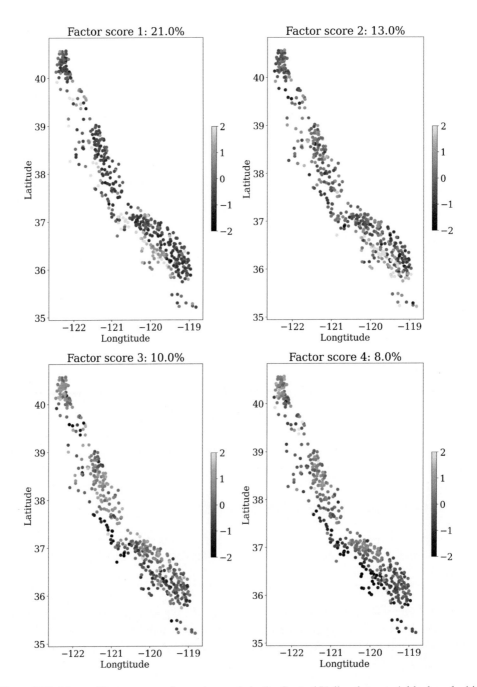

Figure 2.52 Maps of factors scores for each sample in the Central Valley dataset. A black and white version of this figure will appear in some formats. For the color version, refer to the plate section.

next best electron acceptor (generally) and its presence in groundwater indicates that the water is oxygenated, since it is not consumed. Uranium, chromium, and vanadium are all mobile in their oxidized states as oxyanions, so the negative loadings and negative scores suggest more oxic groundwater conditions.

- Factor 2: The presence of major cations of calcium and magnesium, and other ions like nitrate, is likely due to an agriculture cause. Arsenic and iron may be geogenic or human-influenced and are known to co-occur.
- Factors 3 and 4: Sodium and chloride ions can be associated with marine-derived rocks, like sulfate in factor 4; these are present in the Coast Range. Uranium in factor 4 has negative loadings. Factor 4 negative scores indicate higher uranium contamination in the Tulare basin, which is at the southernmost end of the Central Valley.

2.9.6 What Have We Learned in Section 2.9?

- We have learned about FA, which is a statistical method that may aid in understanding causes.
- We have learned that the purpose of FA is very different from PCA. Factor analayis attempts to explain correlation while PCA explains variation in datasets.
- We have learned that it is good practice in FA to work with an expert. Caution must be exercised in interpreting factor scores in terms of the processes that have taken place.

2.10 What Have We Learned in This Chapter?

With extensive data such as the Central Valley dataset, you are likely not going to be able to do all analysis on your own. You should seek collaboration. In such collaboration it is important to be able to communicate. Communication is about trying to understand the goal and purpose as well as communicating what is seen in various plots that are presented. Collaboration teams consist of domain experts and data scientists. The collaboration does not start from any data analysis; first we ask:

- What is the purpose of this study?
- What scientific hypotheses are you interested in investigating?

  The next series of questions may be around the viability of the data:

- How many variables do you have? How many samples? Do samples contain zeros?
- What is the spatial and/or temporal distribution of the data? Does it make sense to mix data from different regions?

Many of the section titles in the chapter have the word "analysis." We have learned that data analysis is key and should precede any predictions. Factor analysis is both an analysis as well as a way of doing unsupervised regression. The factor model allows for an analysis of the data in terms of a limited number of causes and, with the help of the expert, the statistical results can be used to predict causes, and where they occur in space and/or time.

  The main communication tools have we learned are:

- the biplot
- the loading plot
- the scree plot
- the PCA score plot
- the MDS score plot
- the outlier detection plot
- factor or principal component scores on a map

You can test your knowledge by describing to a non-expert in data science what these plots communicate and why they may be relevant for their purpose and/or the hypothesis in question.

TEST YOUR KNOWLEDGE

2.1 Why don't we just look at high chromium concentrations and forget about multivariate analysis?
   a. Because high chromium concentrations may have geogenic or anthropogenic causes
   b. Because there is more information about chromium in a multivariate analysis compared to a univariate analysis
   c. Because chromium has reactions with other elements that may help us in understanding process
2.2 Why is there a need for compositional data analysis?
   a. To deal with the fact that the data are not standardized
   b. To deal with the fact that a closed sum causes artificial correlation
   c. To deal with the fact that compositions have a too low variance
2.3 When is alr preferred over clr?
   a. When an expert provides us information about what ratios to look at
   b. When we have zeros in the data
   c. When we want to remove the closed-sum problem
2.4 What can you observe in a biplot?
   a. Differences in means of subcompositions
   b. Missing data
   c. Grouping in compositional data
   d. Important variables such as relative compositions
   e. Correlation between variables
2.5 What is the purpose of multivariate outlier detection?
   a. To remove any unmasked outliers
   b. To identify samples that deviate from what is expected
   c. To see the difference between errors in data and interesting samples
2.6 What is the main purpose of PCA?
   a. To better understand correlation in multivariate or compositional data
   b. To perform dimension reduction

    c. To remove variables from the dataset

    d. To remove samples from the dataset

2.7 What can a loading plot be used for?

    a. To approach an expert to get a better understanding of the meaning of principal components

    b. To identify the combination of variables that seem to drive correlation in the dataset

    c. To perform dimension reduction

2.8 Why do we need to standardize data before performing PCA?

    a. Because we want to perform an analysis that is not influenced by units

    b. Because it maximizes correlation, hence enhancing the PCA analysis

    c. Because we don't care about the variance of each variable, only about the correlation between variables in finding interesting directions in the data

2.9 What is the practical use of the scree plot?

    a. To determine the information lost in dimension reduction

    b. To determine important variables in our dataset

    c. To assess the variance of the principal components

    d. To assess correlations between the principal components

2.10 What is dual with each other?

    a. MDS and PCA

    b. Variance–covariance of variables and Euclidean distance samples

    c. FA and PCA

2.11 What limitation does PCA have that MDS not have?

    a. In PCA, you are stuck with the correlation coefficient, while in MDS you are not stuck with the Euclidean distance

    b. MDS is computationally less demanding than PCA

    c. MDS works well when you have lots of sample, while PCA does not

2.12 What is a medoid?

    a. It is like a median but in higher dimensions

    b. It is a special type of ellipsoid

    c. In multivariate analysis, it is the middle sample

2.13 What is the most important difference between FA and PCA

    a. FA concerns understanding variance, while PCA is about understanding correlation

    b. FA is about understanding correlation, while PCA is about understanding variance

    c. They have a different purpose: FA is about detecting causes, PCA is about analyzing data

FURTHER READING

Several excellent textbooks on computational data analysis have been produced accessible as introductory texts or software; some examples are:

- Pawlowsky-Glahn, V., Egozcue, J. J., and Tolosana-Delgado, R. (2015). *Modeling and Analysis of Compositional Data*. John Wiley & Sons.
- Filzmoser, P., Hron, K. and Templ, M., 2018. *Applied compositional data analysis*. Springer.
- Van den Boogaart, K. G. and Tolosana-Delgado, R. (2013). *Analyzing compositional data with R*. Springer.

Two textbooks with the same title can be explored for further learning in multivariate analysis:

- Johnson, R. A. and Wichern, D. W. (2014. *Applied Multivariate Statistical Analysis*. Pearson.
- Härdle, W. K. and Simar, L. (2019). *Applied Multivariate Statistical Analysis*. Springer.

There are some papers with a geosciences focus that may help in understanding applications:

- Reimann, C., Filzmoser, P., and Garrett, R. G., 2002. Factor analysis applied to regional geochemical data: problems and possibilities. *Applied Geochemistry*, 17(3), 185–206.
- Filzmoser, P., Garrett, R. G., and Reimann, C. (2005). Multivariate outlier detection in exploration geochemistry. *Computers & Geosciences*, 31(5), 579–587.
- Templ, M., Filzmoser, P., and Reimann, C. (2008). Cluster analysis applied to regional geochemical data: problems and possibilities. *Applied Geochemistry*, 23(8), 2198–2213.

REFERENCES

Dheeru, D. & Casey, G. (2017). UCI Machine Learning Repository. University of California, Irvine, School of Information and Computer Sciences. http://archive.ics.uci.edu/ml

Dickinson, W. R. & Suczek, C. A. (1979). Plate tectonics and sandstone compositions. *AAPG Bulletin*, 63(12), 2164–2182. https://doi.org/10.1306/2F9188FB-16CE-11D7-8645000102C1865D

Egozcue, J. J., Pawlowsky-Glahn, V., Mateu-Figueras, G., & Barcelo-Vidal, C. (2003). Isometric logratio transformations for compositional data analysis. *Mathematical Geology*, 35, 279–300.

Fakhreddine, S., Babbitt, C., Sherris, A., et al. (2019). *Protecting Groundwater Quality in California, Management Considerations for Avoiding Naturally Occurring and Emerging Contaminants*. Environmental Defense Fund. www.edf.org/sites/default/files/documents/groundwater-contaminants-report.pdf

Pawlowsky-Glahn, V., Egozcue, J. J., & Tolosana-Delgado, R. (2011). Lecture Notes on Compositional Data Analysis. www.compositionaldata.com/material/others/Lecture_notes_11.pdf

Poland, J. (1977). Location of maximum land subsidence in US Levels at 1925 and 1977. US Geological Survey. www.usgs.gov/media/images/location-maximum-land-subsidence-us-levels-1925-and-1977

Tarvin, K. (2002). Blue jay measurements. https://search.r-project.org/CRAN/refmans/Stat2Data/html/BlueJays.html

Vigen, T. (2015). Spurious correlations. www.tylervigen.com/spurious-correlations

Wikipedia Commons (2008). Chromite. https://commons.wikimedia.org/wiki/File:Chromit_1.jpg

Wikipedia Commons (2009). Chromium. https://commons.wikimedia.org/wiki/File:Chromium.jpg

Wikimedia Commons (2010). California Central Valley Grasslands map. https://commons.wikimedia.org/wiki/File:California_Central_Valley_Grasslands_map.svg

3   Spatial Data Aggregation

Expected Learning Outcomes

- You will learn that a sustainable energy future will heavily rely on successful mineral exploration.
- You will learn that finding resources in the subsurface is difficult; that many interesting geophysical or geochemical anomalies turn out to be nothing at all: a false positive problem.
- You will learn about Bayes' rule: a general formulation involving making predictions based on what is currently known, and how new information changes predictions by changing the probabilities of what we'd like to predict.
- You will learn about the counterintuitive consequences of Bayes' rule, such as the surprising fact that rare events are harder to predict than most imagine, even with very accurate measurements.
- You will learn about ways to assess the accuracy of predictions made.
- You will learn about an alternative to Bayes' rule, termed logistic regression.
- You will learn to apply all these concepts to a real mineral exploration venture in Northern Quebec.

3.1   Introduction

Humans worldwide collect information about the planet, with various goals in mind: to predict weather, to predict climate change, to quantify hazard occurrence, to determine where natural resources are. Much of this information is spatial and/or spatio-temporal in nature. Remote sensing is one important information source. It allows the inspection of various aspects of the Earth surface or subsurface, or any changes happening there. Various types of remote sensing and geophysical methods exist. For example, in geophysical imaging of the Earth, one often uses a physical source, such as an explosion or an electrical signal, to probe the subsurface. The source creates subsurface responses depending on the nature of those properties; then, these responses are collected in receivers and analyzed. This analysis often results into a map or a three-dimensional "image" of the subsurface in terms of a physical property (e.g., electrical conductivity).

A second way to access information about the Earth is not from a distance, but right here, in or on the ground, such as collecting soil samples, or drilling holes in the ground to observe the subsurface directly. This second type of information gathering leads to much less coverage of the Earth surface or subsurface, but often the information is more aligned with what we want to know: a soil analysis of water content, composition of toxic metals, permeability of a core, grade of nickel in a borehole, etc.

All these datasets are sources of information that need to be aggregated. Aggregation refers here to the way various information sources are combined, usually into a single map, or "putting parts into a whole." Indeed, each information source may provide some part of the puzzle. Data science tools are needed to present a single unified view, from which important decisions can be reached or conclusions drawn.

We focus here on tools that apply well to problems of natural hazards and natural resources. Before thinking about methods that can predict these phenomena, it is useful to think how they will be used once predictions are made. Let's start with landslide predictions. Landslides cause thousands of deaths and billions of dollars of damage per year. They may occur in remote areas, such as Nepal, where access to emergency services is limited. They also occur in urban settings, often with a large number of casualties. Current predictions are that landslides will only increase in intensity and occurrence with climate change, for example because of an increase in extreme rainfall. Rainfall and earthquakes are termed trigger factors. After the trigger factors, the second important part to predicting landslides is the so-called environmental factors. The third part is what elements are at risk, such as people, buildings, roads, etc. The environmental factors are often spatial in nature and can be acquired from remote sensing, for example land use, geological formation, terrain, etc. If we look only at environmental factors, then we want to know which ones are most important. Secondly, we'd like to combine all factors into a single map that expresses a susceptibility to landslides. This map then needs to be combined with the elements at risk, the frequency of landslide occurrence. Many of these ideas are covered in courses and books that focus on landslides. The tools covered in this chapter focus on how we can combine the occurrence of known (historical) landslides, also termed "landslide inventory," to model the current susceptibility of landslides from the environmental factors.

 Visit **Notebook 12: Landslides** to get better insight into datasets used for landslide prediction.

3.2   Exploration for Battery Metals

Hazards are one example of how spatial information needs to be aggregated to plan the distribution of emergency services. A second example is in the exploration of Earth resources, by which we mean oil/gas, heat (geothermal), water, minerals, and storage sites (carbon dioxide

[CO_2], nuclear waste, hydrogen). In this section we use the exploration of battery mineral resources as a demonstration study.

3.2.1 Enabling a Sustainable Future through Batteries

Global development of renewable energy technology and infrastructure will increase the current (2021) demand for metals such as lithium (Li), cobalt (Co), copper (Cu), and nickel (Ni), as well as graphite, by approximately 450% over the next three decades, with an estimated cost of $10 trillion and possibly much more. Fully electrifying the light-duty auto fleet requires the discovery of new deposits, and fast, because each year that passes, internal combustion engines keep adding CO_2 to the atmosphere. Electrification also means that electricity will need to be established from green sources, such as wind and solar energy, which are unfortunately intermittent. Batteries will need to be part of the sustainability outlook for short- to medium-term storage. They can be recycled to a large degree, creating a circular economy once they are mined. However, high-energy batteries that are thermally stable require various kinds of metals such as nickel and cobalt. The infrastructure of electrification and charging stations will require a large amount of copper.

In order to meet this demand, we must be able to efficiently evaluate prospective targets for mineral extraction. Mining companies maintain extensive portfolios of exploration targets and are limited by the rate of evaluating these targets. If additional mineral resources are not discovered, extreme stress will be placed upon existing "artisanal" mining operations, which are unable to extract metals while maintaining stewardship of the environment and human life. Artisanal mining operations employ approximately two-thirds of all miners, and these operations are unable to handle pollution appropriately via heavy metals or ensure proper safety standards. One extreme case is cobalt; about two-thirds of the world's known reserves are located in the Democratic Republic of the Congo (DRC). The problem is that the path from mineral exploration to actual mining is very slow. Hence, new tools are needed to speed up the discovery rate.

3.2.2 Exploring on Cape Smith, Canada

The data covered in this chapter are related to an actual exploration project by a mineral exploration company in the Cape Smith area, northern Canada, near an existing mine, the Raglan mine; see Figure 3.1. All data were acquired in 2021. We are looking for massive sulfide deposits known to host large amounts of nickel, cobalt, and copper; in other words, the metals needed for a sustainable energy future. In this chapter, we'd like you to imagine being an exploration geologist with boots on the ground in the summer of 2021. Alternatively, if preferred, because that area is cold and barren, you can imagine being a data scientist in the office of the exploration company, sitting behind a computer, processing, analyzing, and predicting.

The setting is as follows: prior to getting to Cape Smith the company has used extensive databases of existing nickel deposits, boreholes, as well as geophysical and geochemical data to map where in Canada significant potential exists. Such maps are termed mineral potential

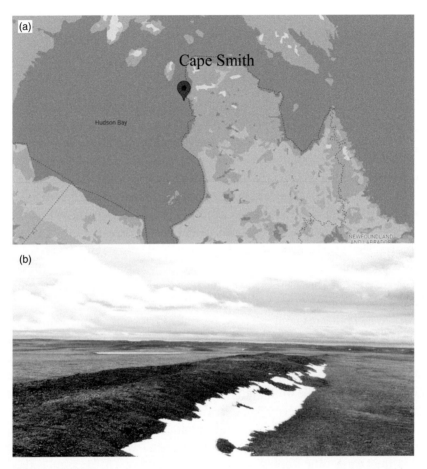

Figure 3.1 (a) Locations on the Cape Smith study area; (b) field outcrop at Cape Smith (courtesy of KoBold Metals®).

maps (MPMs), which are regularly composed by various geological surveys around the world. In fact, for Canada, a similar map was created by the Canadian Geological Survey; see Figure 3.2. The creation of these maps can be achieved with the ideas covered in this chapter. Here we start with an MPM, and focus in on an interesting prospective area in Cape Smith. However, MPMs, or any such maps, including for those detailing hazards, are usually at a fairly low resolution. For example, the Canadian survey map is at 5×5 km resolution. If we are looking for a 100–500-meter large orebody, higher-resolution information is required. Additionally, such information will need to be from the subsurface as well. There are several ways to get this; one is by surface-based geophysical imaging, and another is drilling, although the latter is very expensive. The third option is for geological fieldwork, but that can only happen if important rocks outcrop at the surface and are not hidden under a thick layer of sediment or other uninteresting material. We are in luck at Cape Smith, where the area is very barren, as shown Figure 3.1b, with many rocks outcropping. This means that geological fieldwork may reveal many clues about the existence of deposits. However, given the vastness

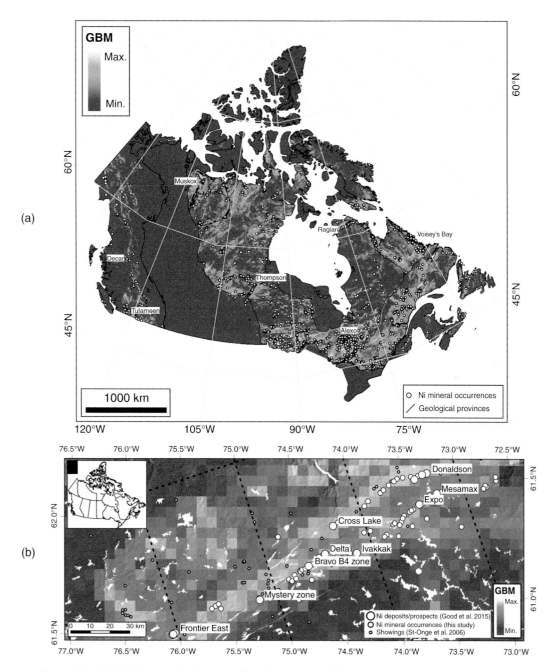

Figure 3.2 (a) MPM created using a gradient boosted machine model (GBM) by the Geological Survey of Canada. (b) More detailed zoom-in of the Cape Smith area. The resolution of this map is 5 x 5 km. The lighter colour indicates high prospectivity. (Figures from Lawley et al., 2021, distributed under a CC BY 4.0 license.) A black and white version of this figure will appear in some formats. For the color version, refer to the plate section.

of this area, doing fieldwork the traditional way may not be effective and requires a lot of time. The modern approach is for such fieldwork to be guided by data science.

3.2.3 How Does Mineral Exploration Work?

To properly use data science to guide fieldwork, we need to understand the fundamentals of mineral exploration and the associated data. We first cover how exploration works, which data are used, and what the particular setting is for Cape Smith. We then present the concepts and tools that will allow for data-science-guided fieldwork.

Understanding Geological Processes

While we focus on one particular area of the world and one type of metal deposit, the same process may be used for exploring for other minerals and indeed other types of resources, such as geothermal energy. A key part in finding mineral deposit systems is to understand how they are formed. In exploration, a mineral system is often the container of rocks that provide the guide to the ore deposit, as the signals in geology, geochemistry, and geophysics are on similar scales as the data. Exploration is divided into "brownfield," exploration near an existing known deposit, sometimes a mine, or "greenfield," where no prior deposits are known.

Geological understanding helps in identifying geological factors (e.g., presence of faults) that are likely to matter, or, even better, to exclude factors that may not matter. Because many massive sulfide deposits exist in all corners of the world, geologist have been able to build, over many decades of work, conceptual models of their genesis. Conceptual models are not like numerical three-dimensional models; rather, these are often drawings illustrating the various processes. They provide a means for a discussion about the processes since the full process may not be exactly understood. After all, we are trying to model something that may have happened hundreds of millions of years ago, starting as deep as the Earth's mantle. In addition, since then, many other processes have taken place, such as folding, faulting, erosion, geochemical alternation, etc.

Like many subsurface resources, the following key elements are needed:

- a source: magmas containing ore-forming elements
- a pathway: magmas need to migrate to the surface
- a trap: a mechanism that can be physical (magma traps) and/or chemical (creating sulfide accumulations).

The process of source, pathway, and trap is illustrated in Figure 3.3, which is specific for deposits such as this one in the Cape Smith belt.

Partial melting (not all rock-forming minerals melt at the same temperature) of the mantle source produces magmas; if the source contains metals, moderate-degree melts can extract them from the source for subsequent enrichment. Mantle material is mafic in composition, which means it is rich in elements such as magnesium and iron, as opposed to felsic (high in feldspar, rich in silicon, depleted in iron and magnesium). The hot magma rapidly spreads below the

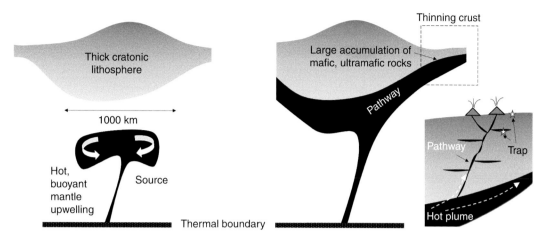

Figure 3.3 Broad overview of the process that creates massive sulfide deposits.

lithosphere to zones where the crust has thinned, for example because of upper-mantle convection or deep-mantle hot-plume activity (Figure 3.3).

Magma can only reach the surface because of openings created by major faults or when cross-linking structures in transform faults offer an open space for magmas to exploit. These then lead to the generation of dikes. Dikes are tabular or sheet-like bodies of magma that cut through and across the layering of adjacent rocks. Eventually, eruption causes lava to flow out onto surface in the direction of least resistance (Figure 3.3). It is important that magmas don't lose their metals too deep below the current level of exhumation of the crust. Deposits can be found in the bottom part of dikes, or lava tubes.

How high concentrations of sulfides are formed, and hence nickel trapped in place, is still less clear. Clearly, there has to be a source of sulfur. Such sources may come from interaction with the crust, or interaction with sulfur-rich mudstone on the surface when the lava mixes with sedimentary rocks after eruptions. Eventually, high concentrations of dense nickel sulfide may settle at the bottom of the lava-tube channels.

Understanding the geological processes is just one element in mineral exploration. It is also important to know where these processes could have taken place. The latter simply means collecting data from fieldwork, geochemical sampling, and geophysical imaging. A geological understanding will help to place all these measurements in a unifying context.

Geophysical Imaging

Rocks in the subsurface contain minerals, and sometimes also fluids in pores (e.g., sandstone containing water). This composition gives rise to various physical properties, such as the density, resistivity, and magnetic susceptibility. Geophysical imaging uses physical signals (e.g., electromagnetic fields, sonic waves, radar waves, etc.) to detect changes in these physical properties. These changes may indicate changes in rock properties, which may then be indicative of the composition of the rock, and hence be used to detect the orebody. This can then be economically exploited. As you can see, that's a long process:

signals $\rightarrow$ physical laws $\rightarrow$ rock properties $\rightarrow$ economic properties

Going backward this way is difficult because different physical properties may yield the same signal, and rocks with different properties share the same physical properties. In mineral exploration for metals, we typically look for magnetic and density anomalies; hence, the measurements used are magnetic and gravity surveys. In Chapter 4, we delve a bit deeper into the use of geophysical surveys in data science for the geosciences.

Geochemical Anomalies

As for the groundwater quality discussed in Chapter 2, geochemical analysis is an important part in mineral exploration. Geochemical samples can be taken in soils easily and collected over a wide area. The hope here is that some kind of footprint exists at the surface of what happened below the surface. This may not always be true. For that reason, surface geochemical sampling is augmented with drill-hole sampling, where the intrusive rocks can be analyzed.

Creating MPMs

Creating MPMs is a typical spatial data aggregation problem. Several sources of information are available to predict the occurrence of a target mineral, often provided in terms of probability. Typically, the following spatial datasets are used:

- geological datasets: rock types, geological ages, faults
- geophysical datasets: magnetic, gravity, electromagnetic data
- geochemical datasets: soil geochemistry, borehole geochemistry
- databases of existing deposits, boreholes with nickel

The first three in this list are the input (predictors), the last one is the output (predictand). The latter are also termed "label data" in machine learning. In this chapter, we go into great detail about the methodologies that use these data to make MPMs in terms of a probability. The map in Figure 3.2 was created by the Canadian Geological Survey. The mineral potential is presented here as a probability. MPMs don't tell you: "if you drill at x, the probability of mineral occurrence is p." Such maps, like for example hazards maps, are guides, not literal probabilities of success or failure. Most importantly, the resolution of these maps is very low relative to what we are looking for, ore bodies. As mentioned above, in the map shown in Figure 3.2, each pixel covers a 5×5 km area. In creating the MPM we compare nickel occurrences, which are usually at a very small scale – a rock at an outcrop or a borehole – with information at much bigger scale, such as geophysics. Hence the MPM misses an important uncertainty: how to compare small scale with large scale, a topic that we also discuss in Chapter 4. MPMs are, however, a good start for targeting actual fieldwork, which is where our story starts.

3.2.4 Data Science Guided Fieldwork at Cape Smith

In the above section, we covered the foundations of mineral exploration and mineral prospectivity mapping. We can now start to plan the Cape Smith fieldwork using data science.

Purpose

A mineral exploration company has created a mineral prospectivity map of Canada and identified Cape Smith as a target for further exploration. The company is also motivated by the existence of a mine (the "Raglan" mine). Cape Smith is far north, and any exploration will need to be done in the summer over a two-month period. The company's plan is to collect more airborne electomagnetic data as well as rock samples. Time is of the essence. The window is short, and weather conditions even in summer are not always optimal for an airborne survey.

The traditional way of mineral exploration is for geologists to do fieldwork on exposed rocks, identify rock types, mineralogy, and structures such as fold and faults. Rock samples are collected and sent to the laboratory for geochemical analysis. Then, usually when field season is over, the results are analyzed, possibly resulting in an additional field campaign planned for the next season. This is a slow process. How can data science help? The traditional approach wouldn't be fast enough to find those metal deposits needed for building the electric vehicle fleet that saves the planet!

Rock-Type Data

The first opportunity for data science is to build a more detailed prediction of rock-type occurrences than the current 5 × 5 km maps. Cape Smith is quite barren, so rocks are exposed at the surface, if there is no snow coverage. That means that remote sensing data, collected in the summer, can be used to quantify the spatial variation of rock types. Cape Smith is a well-studied area in terms of what geological processes created the massive sulfide deposits. The proximity of the mine certainly helped in that regard. What we are looking for are intrusive rocks, according to the schematic explanations in Figure 3.3, in particular, mafic intrusive rocks (gabbro) and ultramafic intrusive rocks (peridotite); see Figure 3.4. The area is also covered extensively in mafic extrusive rocks, in particular basalt. We are looking for intrusions, not extrusive rock formed on the surface of the Earth from lava. We now have two datasets:

- The "labels": known observations of intrusive rocks (peridotite or gabbro) and basalt, from previous fieldwork, as well as from interpretations of the remote sensing data. Here, we label intrusive rock as "1" and basalt as "0."
- The "input": remote sensing data.

Here is the main idea:

- Build a high-resolution map that predicts the presence of these two rock types, with a resolution of 6 meters.

Figure 3.4 Various rock types of interest. There is an extensive cover of basalt in the exploration area, but we are looking for intrusive rocks, including gabboro and peridotite. A black and white version of this figure will appear in some formats. For the color version, refer to the plate section.

- Use this map to plan an optimal path for fieldwork to visit the likely locations of intrusive rocks.
- As the rocks are collected in the field, send the new rock-label data to the data scientists, who then update the predictions, hence allowing an update in the path that the field geologist needs to take.

We will deal with two rock-label datasets:

- The rock-label dataset prior to field data. This will be termed: Day 1 data.
- The rock-label dataset after fieldwork starts, on the twentieth day: Day 20 data.

Of course, the remote sensing data remain the same over the field campaign. Figure 3.5 shows the area of study and the rock-label data collected.

Remote Sensing Data

As the word suggests, in remote sensing we acquire data from a distance. For this Cape Smith study, this means satellite data, in particular WorldView-3 data. Why satellite data? And what kind?

In Cape Smith, we'll use a remote sensing method termed hyperspectral imaging. This type of imaging of the Earth's surface is particularly useful for identifying and characterizing materials, exactly what we are interested in, in this case, the various rock types. A remote sensing dataset is a multivariate dataset; as the name suggests, we are dealing with a spectrum of spectral bands. Information obtained about reflected light energy can be split into different wavelengths, and each of these bands may be indicative of a different element that we are studying. Figure 3.6

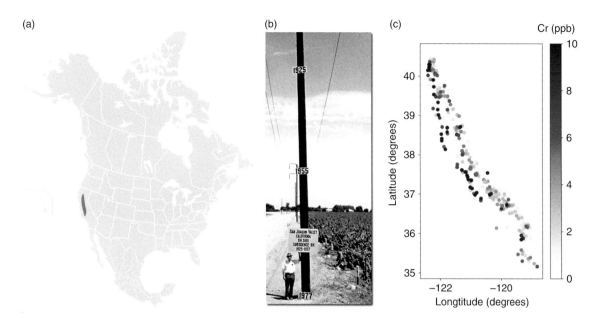

Figure 2.1 (a) Location of the Central Valley of California (California Central Valley Grasslands Map, distributed under a CC BY-SA 3.0 license; Wikimedia Commons, 2010). (b) Subsidence due to overpumping of the groundwater system. The signs on the pole show the approximate altitude of the land surface in 1925, 1955, and 1977. The site is in the San Joaquin Valley southwest of Mendota, California (photographed by Joseph F. Poland, US Geological Survey) (Poland, 1977). (c) Groundwater concentration (parts per billion [ppb]) of Cr from January 2018 to January 2019.

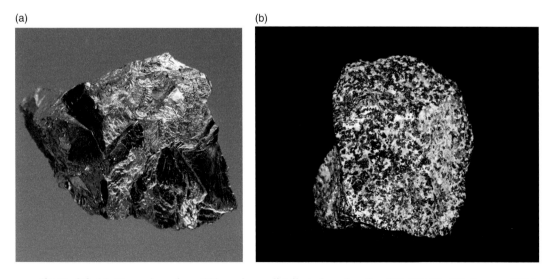

Figure 2.2 (a) Pure chromium (Chromium, distributed under the CC BY-SA 3.0 license; Wikipedia Commons (2008). (b) Chromite ore (Chromite, distributed under the CC BY-SA 4.0 license; Wikipedia Commons (2009).

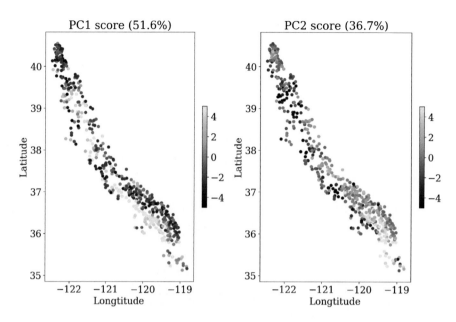

Figure 2.35 Mapping principal component scores onto the sample locations.

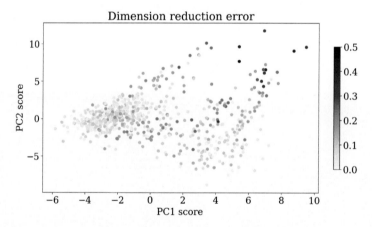

Figure 2.36 Score plot showing the dimension reduction, with the value of the relative errors shaded according to the scale shown.

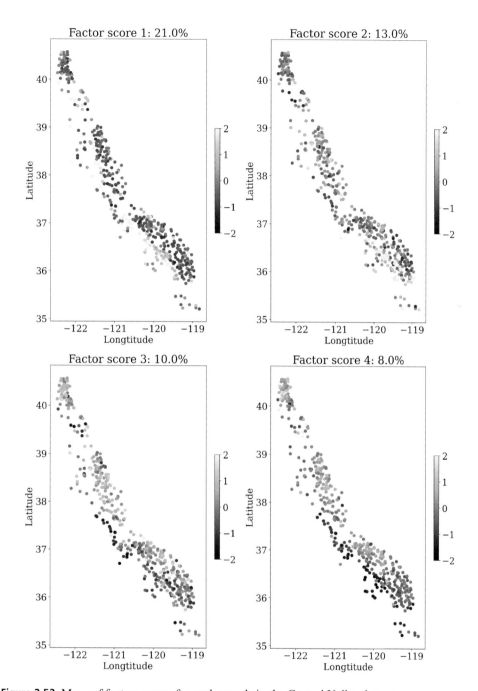

Figure 2.52 Maps of factors scores for each sample in the Central Valley dataset.

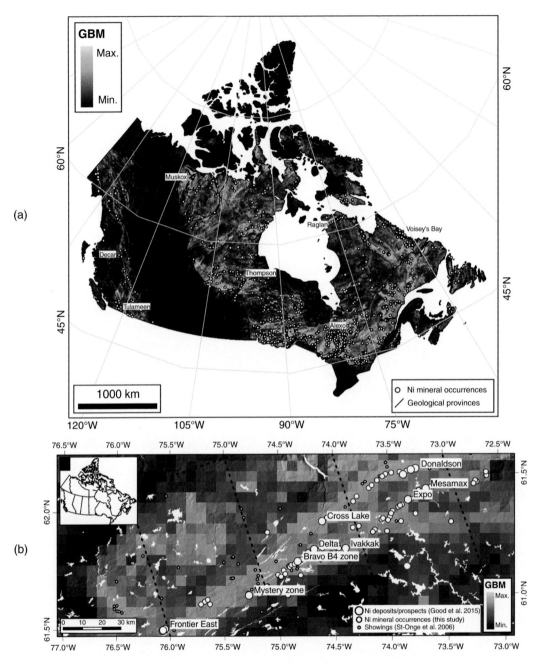

Figure 3.2 (a) MPM created using a gradient boosted machine model (GBM) by the Geological Survey of Canada. (b) More detailed zoom-in of the Cape Smith area. The resolution of this map is 5 × 5 km. The lighter colour indicates high prospectivity. (Figures from Lawley et al., 2021, distributed under a CC BY 4.0 license).

Figure 3.4 Various rock types of interest. There is an extensive cover of basalt in the exploration area, but we are looking for intrusive rocks, including gabboro and peridotite.

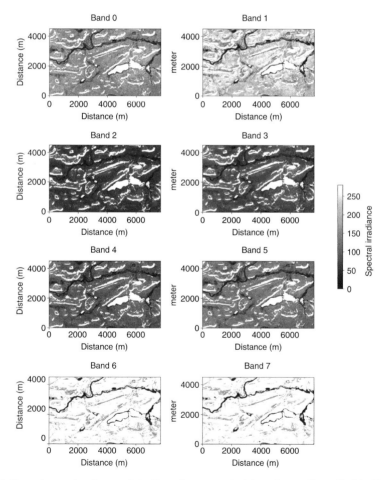

Figure 3.6 Remote sensing image data from hyperspectral imaging at Cape Smith. Each image is from a VNIR spectral band. There is a total of eight bands (see parameters in Table 3.1).

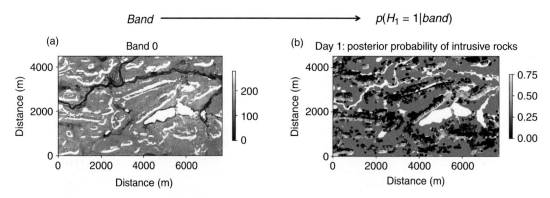

Figure 3.8 (a) Data for the single band 0 are transformed to (b) the probability of intrusive rock, using Bayes' rule. The shaded (colored) dots are the land covered by rocks.

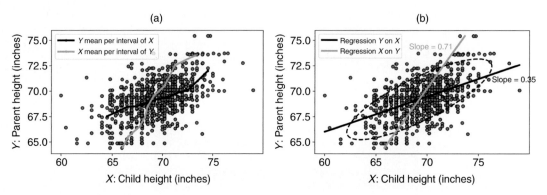

Figure 3.16 Galton's regression data in modern software, showing (a) the conditional mean of $Y|X$, and of $X|Y$; and (b) the linear regression.

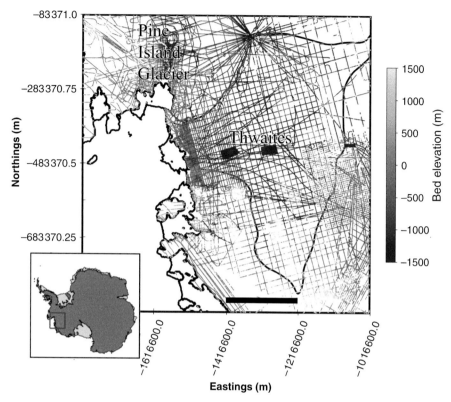

Figure 4.2 Radar flightline data acquired over West Antarctica (from Yin et al, 2022, distributed under a CC BY 4.0 license).

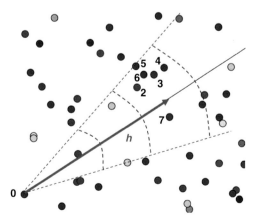

Figure 4.16 Collection of irregularly space samples.

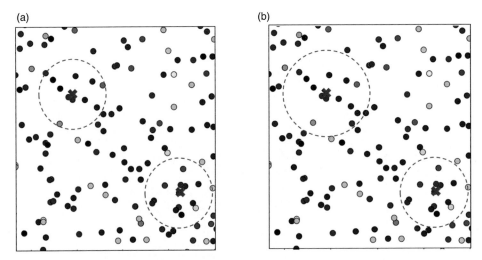

Figure 4.26 Two types of neighborhood definition: (a) with constant radius search neighborhood; (b) with a constant number of samples (14) in the neighborhood.

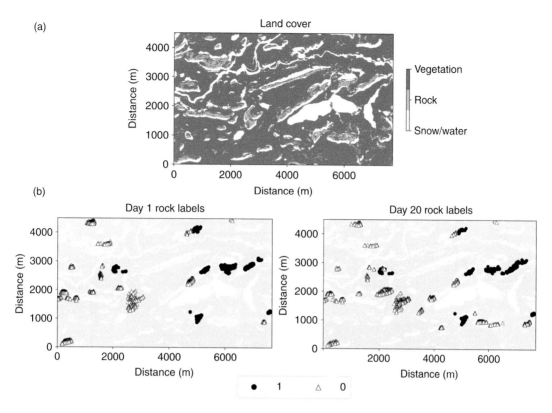

Figure 3.5 (a) Area of study within the Cape Smith belt, showing the land cover. (b) Rock-label data (0, basalt; 1, intrusive rock [peridotite or gabbro]).

shows the available eight visible near infrared (VNIR) spectral bands. They will be used as the predictors of rock type.

 Visit **Notebook 13: Cape Smith Mineral Exploration Data** that provides you with an overview of all the data.

3.3 Concept Review

In this chapter, we cover important concepts such as conditional probability and Bayes' rule. Section 5.4 provides a refresher as well as a deeper dive into:

- the rules of probability
- where Bayes' rule comes from and the motivation behind it
- the binomial distribution, which plays an essential role in calculating the chance of events or the combination of events
- the concept of "maximum likelihood," which is fundamental to statistical inference

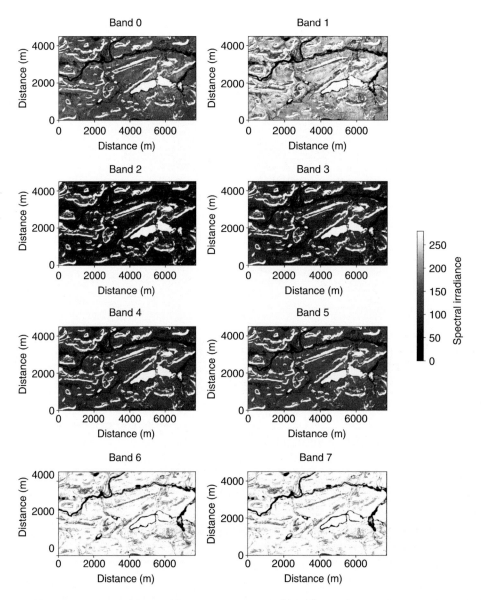

Figure 3.6 Remote sensing image data from hyperspectral imaging at Cape Smith. Each image is from a VNIR spectral band. There is a total of eight bands (see parameters in Table 3.1). A black and white version of this figure will appear in some formats. For the color version, refer to the plate section.

3.4   Predicting with Bayes' Rule

The goal of the Cape Smith study is to make predictions of occurrence of rocks over some area considered of interest. What is a prediction? It is a word that is used very loosely but also has a very precise meaning in data science. If we ask a meteorologist to "predict the weather," they will provide statements of certainty and statements of uncertainty. A prediction can be like

Table 3.1 Wavelength ranges of the spectral bands, from the available remote sensing data at Cape Smith

	Band name	Wavelength range (nm)		Band name	Wavelength range (nm)
Band 0	Red	630–690	Band 4	Green	510–580
Band 1	Near-red edge	705–745	Band 5	Yellow	585–625
Band 2	Coastal	400–450	Band 6	Near-infrared 1	770–895
Band 3	Blue	450–510	Band 7	Near-infrared 2	860–1040

that: we state what we know (it will be cloudy) and what we don't know fully (the amount of rain). Let's focus on "don't know," or lack of knowledge. How do we make a statement of such uncertainty? There are several ways to approach this, including:

- making a guess and then saying how much confidence one has about that guess
- making a statement of chance or probability

For example, the meteorologist may state: "there will be 1 inch of rain in the next 24 hours" and provide some $+/-$ around that statement. Or they can make a probability statement: the chance of getting more than 1 inch is 50%. The latter statement is, however, more general. In fact, if we make that statement not just for 1 inch, but for 0.2, 0.4, 0.6, etc., we have the full uncertainty quantified as a cumulative probability function. The first statement remains incomplete, the $+/-$ remains indicative; it does not say that 1.2 is the high end and 0.8 the low end with certainty. It is fuzzier than this.

Statements of uncertainty that are based on a probabilistic expression are therefore the more general way of expressing uncertainty; see Section 5.4. They are based on information. The meteorologist does not make random guesses; in fact, their prediction may be based on quite a lot of information. Information may also change the uncertainty statement. The statements on rain may be refined as more information becomes available; eventually, it may either become 0 (no rain for sure) or 1 (rain for sure). The way in which this change of probability works follows a rule: Bayes' rule.

3.4.1 Bayes' Rule

Meteorologists make predictions based on information, and geologists don't just wander around randomly. If we have a lot of informative data (i.e., data that actually tell us something), the uncertainty will be smaller, meaning the probabilities become closer to 1 or 0. So, what we need is something that does this quantitatively, a function that tells us what the probability of some event of interest is, given the amount of informative data we have. To introduce this function, let's start with a game and a problem based on that game.

Monte Hall Game

This problem is inspired by an American television game show, *Let's Make a Deal!*, hosted by the presenter Monty Hall. On the show, a guest is presented with three closed doors. Behind one door is a car; behind the others, goats. Imagine the guest picks door 3. Then, to raise suspense, the host asks, "Wait, are you sure?" "Hold on, I am going to open a door for you." The host, who knows which door the car is behind, then opens another door, say door 1, which happens to contain goat. The guest is then asked by the host: "Do you want to stay with door 3 or do you want to switch to door 2?" Then of course he calls on the audience to chime in, raising the suspense even more. Some guests may feel this is a trick, most people may think it doesn't matter: you are left with two doors, so the chance is 1/2, or is it? *It isn't*, and we can know this from a simple rule (Bayes' rule) of probability. But before introducing that rule, we need a way to quantify how information changes knowledge about something we don't know.

Conditional Probability

Information does not change the current yet unknown fact that the car is behind a certain door. Information changes our uncertainty, as measured through probability. However, here we are dealing with a different kind of probability, namely conditional probability.

Our game problem can be mathematically formulated using a binary random variable, or "it is or isn't," which are mutually exclusive. This exclusion is important; we can't say that there is probability that it *may* be behind door 1. If we do that, we are making two vague statements in one sentence: "probability" and "may," and this, according to the rules and axioms of probability, is not allowed (see Section 5.4). Also, we'd like to be able to calculate things, so we need a way to turn "car behind door 2" into an actual numerical value. A simple way to do this is as follows: we can define a variable I_1 as follows: $I_1 = 1$ if the car is behind door 1, and zero if the car is behind doors 2 or 3.

$$I_1 = \begin{cases} 1 \text{ if car is behind door 1} \\ 0 \text{ else} \end{cases} \tag{3.1}$$

where I_1 is commonly known as an "indicator variable." Clearly, we do not know I_1; if we knew $I_1 = 1$, then we have the answer. If we know $I_1 = 0$, we are still uncertain whether the car is behind door 2 or door 3. We can now turn words into quantification:

$P(I_1 = 1) \leftrightarrow$ the probability that $I_1 = 1 \leftrightarrow$ "The probability that the car is behind door 1"

Likewise, we can define I_2 and I_3. So, what we have done is turn a "word problem" into a mathematical formulation. This is very important, and it is important to make sure to get the details correct. What is not correct is to write $P(I_1)$ instead of $P(I_1 = 1)$, although often we use that as short-hand notation. For a novice, this is not recommended. Another way of doing the same thing is using a hypothesis:

H_1: my hypothesis is that the car is behind door 1

This hypothesis is either true (1) or false (0) so we can also write:

$$P(H_1 = \text{true}) \leftrightarrow \text{``The probability that the car is behind door 1''}$$

Similarly, we can define I_2 and I_3: H_2 and H_3.

A first rule is that the car is either behind door 1, 2, or 3, or mathematically:

$$P(I_1 = 1) + P(I_2 = 1) + P(I_3 = 1) = 1 \tag{3.2}$$

This means that

$$P(I_2 = 1) + P(I_3 = 1) = 1 - P(I_1 = 1) \tag{3.3}$$

or

$$P(I_1 = 0) = 1 - P(I_1 = 1) \tag{3.4}$$

Now comes an important element, and that is: information. The host knows (is informed) about the location of the car; hence, in opening a door, they use that information. We need another notation for this information. What we do is to introduce a new binary variable; let's call this variable E (as in evidence)

$$E_1 = \begin{cases} 1 \text{ if host opens door 1} \\ 0 \text{ else} \end{cases} \tag{3.5}$$

Likewise, we have E_2 and E_3, all of which are mutually exclusive. Let's now return to our situation: the guest chose door 3, the host opens door 1. This means that: "If the car is behind door 2, the chance that the host opens door 2 equals zero." In probability terms we write this as follows:

$$P(\text{``host opens door 2''} \mid \text{``the car is behind door 2''}) = 0$$

Why? The host will never open that door, otherwise the prize will be revealed! The "|" is notation for "given the information that" or "given that."

Let's take another situation: you pick door 1, the car is behind door 2, then the host has to open door 3, so

$$P(\text{``host opens door 3''} \mid \text{``the car is behind door 2''}) = 1$$

The final situation is: you pick door 1, and the car is behind door 2, then the host has now a choice, open 1 or 3, we can write that as

$$P(\text{``host opens door 1''} \mid \text{``the car is behind door 2''}) = 0.5$$

$$P(\text{``host opens door 3''} \mid \text{``the car is behind door 2''}) = 0.5$$

Now one may think the problem is solved: the chances are 50/50! Not yet. The probability we need is the following, after you pick door 1, and the host opens door 2:

$$P(\text{``the car is behind door 1''} \mid \text{``the host opens door 2''}) = ?$$

$$P(\text{"the car is behind door 3"} \mid \text{"the host opens door 2"}) = ?$$

So, the question is this: if we know what a probability of the type $P(E|I)$ is, what is $P(I|E)$? We know for sure that in general:

$$P(E|I) \neq P(I|E) \tag{3.6}$$

The answer lies in Bayes' rule.

Deriving Bayes' Rule

In Section 5.4, we describe in detail the story behind Bayes' rule, named after Thomas Bayes, a reverend who lived in the eighteenth century. There is a simple logic behind Bayes' rule, but to get there, we need to introduce another concept, namely joint probability. An example is

$$P(\text{"the car is behind door 3" and "the host opens door 2"})$$

Notice that the "and" is very different from the "|"; the latter indicated a condition. Joint probabilities can be calculated too. To see how, we can turn the "and" back into a "|" as follows:

$$P(\text{"the car is behind door 3" and "the host opens door 2"})$$
$$= P(\text{"the car is behind door 3"})$$
$$\times P(\text{"the host opens door 2"} \mid \text{"the car is behind door 3"})$$

or, without words

$$P(I_3 = 1 \cap E_2 = 1) = P(I_3 = 1) \times P(E_2 = 1 | I_3 = 1) \tag{3.7}$$

The logic here is that for both I_3 and E_2 to happen, I_3 has to happen and then given I_3 happens, E_2 has to happen as well. Now, because "and" is symmetric in its argument, we can also write

$$P(I_3 = 1 \cap E_2 = 1) = P(E_2 = 1) \times P(I_3 = 1 | E_2 = 1) \tag{3.8}$$

If we divide these two equations, then we get rid of the joint probability and we can arrange everything as follows:

$$P(I_3 = 1 | E_2 = 1) = \frac{P(E_2 = 1 | I_3 = 1) \times P(I_3 = 1)}{P(E_2 = 1)} \tag{3.9}$$

This is Bayes' rule. It is often written as follows:

$$P(H|E) = \frac{P(E|H) \times P(H)}{P(E)} \tag{3.10}$$

It is, however, preferrable to write "$E = i$" over omitting the "$= i$".

The Intricate Consequences of Bayes' Rule

Now, we dive a bit deeper into the consequences and meanings behind this simple rule. Imagine a case where you have two scientific theories for the same phenomenon, but 90% of researchers put belief in theory 1 and 10% put belief in an alternative theory 2. So

- H_1: theory 1 is true
- H_2: theory 2 is true, hence H_1 is false

"Belief," as a subjective notion of opinion, is turned into probability:

$$P(H_1 = \text{true}) = 90\%$$

$$P(H_1 = \text{false}) = 10\%$$

In Bayes' rule we call these: *prior probabilities*. The prior refers to the current state of knowledge, prior to doing more experiments or tests.

Of course, scientists don't stop there and do more experimentation, such as laboratory, computational modeling, or field studies. These studies provide additional information or evidence. Let's now consider that a "test" has been done to study these theories and that the outcome of that test is also binary: "positive" $E = 1$ and "negative" $E = 0$, as in medical test results. A positive test result is an indication that H_1 is true. However, no test is perfect, we need to specify a test accuracy. The way scientists do this is by testing the test. They instantiate various conditions where they know theory 1 is true, then see how well the test performs. The accuracy is therefore measured by the following probability:

$$P(\text{test is positive} \mid \text{theory 1 is true})$$

But that's not all, they would also need to specify:

$$P(\text{test is positive} \mid \text{theory 1 is false})$$

A test may be good at the combination of (test positive, theory 1 is true) and not as accurate in the combination of (test positive, theory 1 is false). The conditional probabilities are termed likelihood probabilities.

In this setting, we can choose two types of tests. We can do a test that aims at confirming the leading theory or we can do a test that helps to debunk the leading theory. In the first case, we would use a test that is accurate in terms of detecting the leading theory. We choose a test such that $P(E = 1|H_1 = 1)$ is high. If we want to debunk the leading theory, then we'd choose a test that is accurate in detecting the alternative theory, or we'd choose a high $P(E = 0|H_1 = 0)$.

If we inspect Bayes' rule, we are still missing one probability and that is $P(E)$ or $P(E = 1)$. This bottom term exists because $P(H|E)$ needs to be a valid probability, meaning:

$$P(H_1 = 1|E) + P(H_1 = 0|E) = 1 \tag{3.11}$$

This leads to the following equation:

$$P(E = 1) = P(E = 1|H_1 = 1)P(H_1 = 1) + P(E = 1|H_1 = 0)P(H_1 = 0) \qquad (3.12)$$

Let's do a few example calculations: one with a confirming test, one with a debunking test. Confirming test, high $P(E = 1|H = 1)$:

- $P(H = 1) = 0.9$
- $P(E = 1|H = 1) = 0.9; P(E = 1|H = 0) = 0.5$
- $P(E = 1) = 0.9 \times 0.9 + 0.5 \times 0.1 = 0.86$
- $P(H = 1|E = 1) = 0.9 \times 0.9/0.86 = 0.94$

Debunking test, high $P(E = 0|H = 0)$:

- $P(H = 1) = 0.9$
- $P(E = 0|H = 0) = 0.9 \rightarrow P(E = 1|H = 0) = 0.1; P(E = 1|H = 1) = 0.5$
- $P(E = 1) = 0.5 \times 0.9 + 0.1 \times 0.1 = 0.46$
- $P(H = 1|E = 1) = 0.5 \times 0.9/0.46 = 0.97$

What do we observe? First, in the confirming test, positivity of the test result is very likely, namely 0.86, while in the debunking test it is less likely, at 0.46. However, with the debunking test we get a much greater increase in the posterior probability $P(H = 1|E = 1)$ going all the way to 0.97. Low likely evidence has more power than high likely evidence. So, yes, trying to debunk a theory is more powerful than trying to confirm one. Bayes' rule explains this very clearly.

A second important example producing counterintuitive results is when we predict rare events. Rare events have very small probabilities of occurrence. For example, in mineral exploration, finding actual economical deposits (D) is challenging, and it is sometimes a matter of luck. A good success rate is about 1%. Imagine we now enter a region where we'd like to find minerals. This means we start from an a priori success rate, or:

$$P \text{ (finding a deposit)} = P(D = 1) = 0.01$$

Now, we are given a metal detector. The company selling this metal detector has been thoroughly studying it, testing it on known deposits, and on areas without deposits. They found that the metal detector is very accurate, namely:

$$P(\text{detector beeping} \mid \text{deposit is present}) = 0.95$$

We must be careful here. What that doesn't mean is that when the detector beeps, the chance of a deposit is 95%, indeed, just like Monte Hall:

$$P(E|D) \neq P(D|E) \qquad (3.13)$$

Similarly, their testing of the device found that:

$$P(\text{detector beeping}|\text{deposit is not present}) = 0.2$$

We now have all the ingredients for using Bayes' rule:

- Prior: $P(D = 1) = 0.01$
- Likelihood : $P(E = 1|D = 1) = 0.95; P(E = 1|D = 0) = 0.2$
- Total: $P(E = 1) = 0.95 \times 0.01 + 0.2 \times 0.99 = 0.2075$
- Posterior: $P(D = 1|E = 1) = 0.95 \times 0.01/0.2075 = 0.045;$
  ○ Therefore: $P(D = 0|E = 1) = 0.955$

The counterintuitive result is that, even when the detectors beeps, there is only a 4.5% chance that there is a deposit. The reason for this is the rarity of the event. Rare events are very hard to predict even with very accurate devices, resulting in very high false positives. Bayes' rule shows quite beautifully that predictions depend on two very important concepts:

- the rarity of the event
- the accuracy of the information

$P(H = 0|E = 1)$ is also termed the false positive (FP) rate. It tells us at what rate (frequency/ probability) we will get beeping when there is nothing there. At least in mineral exploration, it is very important to know what this rate is, and also what information we can use to reduce it. Similarly, we have:

- true positive (TP): $P(H = 1|E = 1)$
- true negative (TN): $P(H = 0|E = 0)$
- false negative (FN): $P(H = 1|E = 0)$

Solving the Monte Hall Problem

It's been a long time coming but, using Bayes' rule we can also solve the Monte Hall problem; recall our notation:

$$E_i = \text{``Host opens door } i\text{''}$$

$$H_i = \text{``Car is behind door } i\text{''}$$

You pick door 1, so after opening either door 2 or door 3, we need to assess the following probability: $P(H_1|E_2)$ or $P(H_1|E_3)$. Let's do the Bayes' calculation protocol. In the case when the host opens door 2, and you have a choice to switch to door 3:

- Prior $P(H_i) = 1/3$
- Likelihood: $P(E_2|H_1) = 1/2; P(E_2|H_2) = 0; P(E_2|H_3) = 1$
- Total: $P(E_2) = P(E_2|H_1)P(H_1) + P(E_2|H_2)P(H_2) + P(E_2|H_3)P(H_3) = 1/2 \times 1/3 + 0 \times 1/3$
  $+1 \times 1/3 = 1/2$
- Posterior $P(H_1|E_2) = \dfrac{1/2 \times 1/3}{1/2} = 1/3$
- Compliment: $P(H_3|E_2) = 1 - 1/3 = 2/3$

Answer: you need to switch doors; your chances will double from 1/3 to 2/3!

Value of Information

Value of information (VOI) is a concept used in decision science to quantify the dollar value of information prior to gathering it. This can be very helpful in real-world data-gathering campaigns. For example, mineral exploration companies need to make choices on which data to acquire and where to acquire that information. Having some quantitative metrics will help. VOI is one such metric.

When is information valuable? In VOI we state three conditions:

- Information is valuable when it tells us something more than we already know.
- Information is valuable when it reveals something about a property of interest.
- Information is valuable when we may alter an important decision we'd like to make.

Let's give this some concrete context in relation to exploration. Information is likely to be valuable early in the exploration stage when little is known, and we are very uncertain. This means that the data need to be measured against what is known a priori. We need to use a prior distribution, as is done in Bayes' rule. Information is valuable when it informs something relevant to our decision. You can always take information, for example, knowing where moose roam in Cape Smith does not inform us of important properties as knowing the lithology would. Also, not all information is equally accurate in terms of what it reveals about that interesting property. Geophysical data cover a large volume but don't directly measure what we want, while drilling does, for example. Lastly, there is no value in information if there is no chance that this information will change a decision we are going to make. This is particularly relevant in exploration. We need to decide, at some point, that the area is very interesting, and go ahead with appraisal operations for a deposit; or we just walk away from it. If information cannot discriminate the walk-away versus go-ahead decision, then it is not valuable.

All of this can be made quantitative using Bayes' rule and we can organize our calculation in what is termed a decision tree. Let's do a simple example illustrating the three principles to actually make a quantitative dollar assessment. We want to know if a "metal deposit is present": our target event. We use the following notation:

$$A = \text{"deposit is present"}$$

$$\overline{A} = \text{"deposit is not present"}$$

A priori, probabilities are very low (it is a rare event), as was discussed before; we take $P(A) = 0.01$. If a metal deposit is present, we'll go ahead and mine. If not, we walk away. In assessing a value, we need actual dollar values, such as how many billion a future mine would have as revenue. These values themselves are of course uncertain, so we'll skip that issue here. Instead, let's simply say that we have a value of "100" if we mine and a metal deposit is present, and "0" if we walk away. There is also a cost associated with going ahead when nothing is there, we will use "1" for this, much smaller than "100." We can present the situation in what is called a decision tree. Using this decision tree, we can calculate the expected value of our current project:

$$\text{Prior expected value} = (100 - 1) \times P(A) + (-1) \times P(\overline{A}) = 99 * 0.01 - 0.99 = 0$$

Our current project value is zero because we have a very low probability of success. One way to change low probabilities is to gather more information. Indeed, Bayes' rule tells us that information can change probabilities. Let's consider we are offered a metal detector, so we now have a third choice, "gather information." The detector is not a perfect device. It comes with a statement of accuracy as mentioned above. This statement is:

$$P(\text{detector beeps} \mid \text{deposit is present}) = 95\%$$

$$P(\text{detector beeps} \mid \text{deposit is not present}) = 10\%$$

We are planning to gather information, which is an action in the future. For that reason, we don't yet know its outcome, it could beep, or not beep; this probability comes from the rule of total probability. We also introduce the following notation:

$$B = \text{"detector beeps"}$$

$$\overline{B} = \text{"detector not beeps"}$$

$$P(B) = P(B|A)P(A) + P(B|\overline{A})P(\overline{A}) = 0.95 \times 0.01 + 0.10 \times 0.99 = 0.1085 \tag{3.14}$$

We now have all the ingredients to apply Bayes' rule, A = metal deposits present, $\overline{A}$ = no metal deposits present; B = detector beeps, $\overline{B}$ = detector does not beep:

- Prior $P(A) = 0.01$
- Likelihood: $P(B|A) = 0.95; P(B|\overline{A}) = 0.10$
- Total: $P(B) = 0.1085; P(\overline{B}) = 0.8915$
- Posterior $P(A|B) = 0.95 \times 0.01/0.1085 = 0.0875; P(\overline{A}|B) = 0.9125$
- Posterior $P(A|\overline{B}) = 0.05 \times 0.01/0.8915 = 0.00056; P(\overline{A}|\overline{B}) = 0.9994$

We can now complete the decision tree (Figure 3.7). In the information-gathering branch, let's walk through the future event that may take place:

- In the future, if you use the metal detector, that tool may beep or not, we know the probabilities.

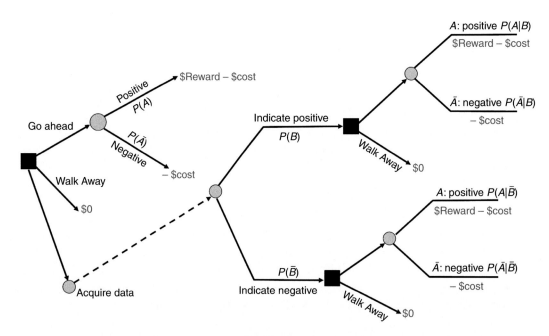

Figure 3.7 Value of information decision tree for a case with a binary decision (go ahead versus walk away), with a binary decision variable *A* and a binary data outcome *B*.

- *After* that happens, we have to make a decision: "go ahead" or "walk away"; it is the same decision as before.
- However, what has changed is the probability of having a deposit, because now we have information, so the *P*("deposits") has changed to *P*("deposit" | "information").

Here are the calculations:

If the future test is positive (*B*):

$$\text{Value}(B) = (100 - 1) \times 0.0875 + (-1) \times 0.9125 = 7.75$$

If the future test is negative ($\overline{B}$):

$$\text{Value}(\overline{B}) = (100 - 1) \times 0.00056 + (-1) \times 0.9994 = -0.94$$

Now add up the value of each branch:

$$\text{Value} = \text{Value}(B) * P(B) + \text{Value}(\overline{B}) * P(\overline{B}) = 7.75 \times 0.1085 + (-0.94) \times 0.8915 = 0.002865$$

So why is this information valuable? Without it, the value of our project is zero. The value is very small, however. We would not expect that, by taking one measurement, a large degree of certainty would emerge. Indeed, we went from 1% to 8.75%. To unmask rare events, we will need to take many more measurements!

Comparing Probabilities

You will never make a mistake when you take two probabilities and multiply them in the sense that a multiplication of a value between 0 and 1 with another value between 0 and 1 is still a value between 0 and 1. That's why in Bayes' rule we always do multiplications.

This is not true for addition, adding a value between 0 and 1 with another value between 0 and 1 is no longer a value between 0 and 1. Another problem is how to compare probabilities. For example, the difference between 0.4 and 0.6 is 0.2 but so is the difference between 0.7 and 0.9; is that the same difference? Subtracting is like adding: it has the same problem here. A simple way around this problem is to use the odds and logarithms of the odds.

The problem is that we are dealing with a bounded variable, between 0 and 1. In addition, if there are more than two categories, probabilities need to sum to one. There are too many constraints to use the usual procedures. A simple way to fix this problem is what we covered in Chapter 2, the log ratios of compositions. Why? A probability is a composition! Log ratios add to a constant and they are non-negative. For a binary event, we have a two-part composition $(p, 1 - p)$, the log odds ratio is nothing more than the additive log ratio of probabilities.

Here is the procedure to transform probabilities to log odds ratios (logit):

- Turn probabilities p into odds: $p/(1 - p)$,
  - $0.2 \rightarrow 0.25; 0.1 \rightarrow 0.11$
- Calculate the logarithm of the odds: $\text{logit}(p) = \log\left(\dfrac{p}{1 - p}\right)$
  - $0.2 \rightarrow -1.38; 0.1 \rightarrow -2.20$

What we have done here is to take a value that is between 0 and 1 (a probability) and turn it into a value between $-\infty$ and ∞ (a log odds ratio). Now we are free to do what we want; we can calculate means and variance as we wish. A very nice property of log odds is that we can invert them back into probabilities as follows; if x is the log odd ratio, the corresponding probability is:

$$p = \frac{1}{1 + \exp(-x)} \tag{3.15}$$

This function is also called the logistic function. Let's return to our question: What is the difference between 0.4 and 0.6 compared to the difference between 0.7 and 0.9?

To solve this, we calculate log odds ratios of 0.4 and 0.6, and only then take the difference: 0.81; for 0.7 and 0.9 this value is 1.35, so they are different! In fact, we can take the log odds ratios 0.81 and 1.35 and transform them back to probabilities, and find them to be 0.69 versus 0.79. The answer therefore is that 0.7 and 0.9 are more different than 0.4 and 0.6. In a way, this also makes intuitive sense: both 0.4 and 0.6 are near 0.5, which means we are very uncertain. However, 0.9 is closer to one, so to go from 0.7 to 0.9 is more difficult, we need stronger information to do so.

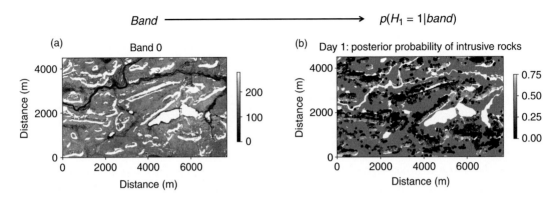

Figure 3.8 (a) Data for the single band 0 are transformed to (b) the probability of intrusive rock, using Bayes' rule. The shaded (colored) dots are the land covered by rocks. A black and white version of this figure will appear in some formats. For the color version, refer to the plate section.

Bayes' Rule Applied to the Cape Smith Data

We illustrate how Bayes' rule can be used on the Cape Smith data in the case of a single band as evidence. Figure 3.8a shows the data at hand, the single band, plus the label (rock-type) data (see Figure 3.5) for day 1 (start of fieldwork) and day 20.

> 💻 Visit **Notebook 14: Bayes** to execute Bayes' rule where the data are only one band of the remote sensing data for Cape Smith.

At each location of the remote sensing data, we have a single observation, let's call it x, the predictor. X is a continuous variable (the remote sensing band), hence the situation is different from previous sections when we just had to count (discrete variables). Bayes' rule, however, remains exactly the same; now, we replace probabilities with density functions (see Section 5.2.2). To understand this, let's look at the likelihood probability $P(E|H)$. The H terms here are the two rock types, so we can turn that into two indicator variables

$$H_1 = \begin{cases} 1 & \text{if intrusive rock is present} \\ 0 & \text{else} \end{cases}$$

$$H_2 = \begin{cases} 1 & \text{if basalt is present} \\ 0 & \text{else} \end{cases} \tag{3.16}$$

Since we have only two rock types, it is sufficient to work only with H_1. First, what is the a priori probability $P(H_1 = 1)$? This is a tricky question. One option is to calculate the proportion based on the label data we have. However, that proportion is certainly biased. Geologists are looking for intrusive rocks, so they will be motivated to create many of those labels. This labeling bias is a very common problem in the geosciences. In fact, the geologist could have decided just to label intrusive rocks, and hence the answer would be 100% intrusive rocks! In this setting, we therefore deliberately take a prior probability that is not equal to the proportion of labels, namely 20% (a rough estimate based on Cape Smith's field geologist). We

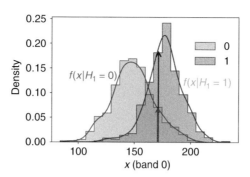

Figure 3.9 Likelihood distributions of the band data for each rock type. The smooth function is obtained using kernel density smoothing (0, basalt; 1, intrusive rock).

will return to this issue later, and investigate what the sensitivity of this value is to the answers we are seeking.

Now, we look at the values of x = band 0 (red) at each location that co-occurs with observations of rock types. We can plot the value of x in two histograms, one for each rock type, as shown in Figure 3.9. In this Figure we observe that the two histograms are not fully overlapping. This means that the x has something to say about rock type, but not perfectly (otherwise there would be no overlap). To apply Bayes' rule, we smooth the histogram. For this we use kernel density estimation (see Section 5.2.2). After kernel smoothing, we know the density value $f(x|H_1 = 1)$ and $f(x|H_1 = 0)$. Now we can calculate the posterior probability. $P(H_1 = 1|x)$ using our Bayes' rule protocol:

- Prior: $P(H_1 = 1) = 0.2$
- Likelihood: $f(x|H_1 = 1)$
- Total: $f(x|H_1 = 1)P(H_1 = 1) + f(x|H_1 = 0)P(H_1 = 0)$
- Posterior:

$$P(H_1 = 1|x) = \frac{f(x|H_1 = 1) \ P(H_1 = 1)}{f(x|H_1 = 1)P(H_1 = 1) + f(x|H_1 = 0)P(H_1 = 0)} \tag{3.17}$$

$$P(H_1 = 1|x) + P(H_1 = 0|x) = 1 \forall x \tag{3.18}$$

What Bayes' rule provides is a mapping of the value of x into a probability value that is function of x: $x \mapsto P(H_1 = 1|x)$. We can apply this function everywhere over the prospective area where we have an observation over x; see Figure 3.8b.

To provide further insight into this, we perform classification, in which we turn a probability of a rock into an actual rock. To do so, we bet on the rock that has the highest probability of being there. This way of classification is also termed Bayesian classification. Accordingly, for each location we pick the rock that has the highest probability; see Figure 3.10. Since only two rock types are considered, the "winner" is that with a probability higher than 0.5. Be careful: this map is not real. It does not mean that when the color is black, you have black for sure. Errors are being made, but we have ways to evaluate the errors and, hence, how much we trust this map.

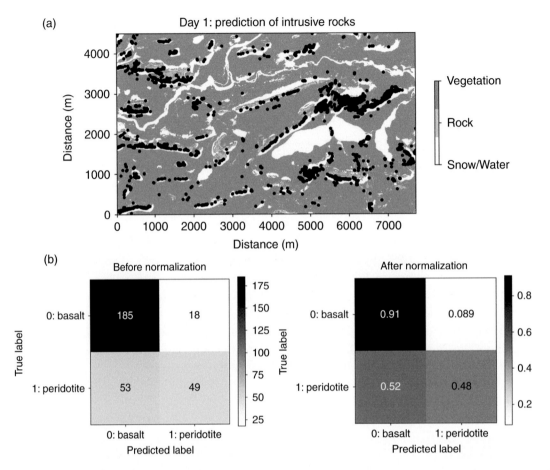

Figure 3.10 (a) Map with Bayesian classification. Classified intrusive rocks are represented by black dots; (b) Confusion matrix (prior = 0.2, threshold = 0.5).

Evaluating Classifiers

A Bayesian classifier takes probabilities and returns a label (rock type). That single label may not be equal to the true label. For that reason, we need to quantify how much error is being made. What we know are the observations at the locations we gathered data. We need to be careful here about the labels: on day 1 all labels were interpretations by a geologist on remote sensing data, so they are in fact not "true labels"; some may be true but some not. However, after day 20, we have collected actual true labels, because geologists went into the field and observed the rock types. For day 1 we have 1522 label data, for day 20 we have 1769 observations (some of the day 1 label data are not included in the day 20 dataset).

A first measure of accuracy is to make a comparison between the true labels and the predicted labels and simply count how many times we performed a correct classification and how many times it was incorrect. Another important aspect of assessing classifiers is to divide the samples into two sets: the "training set" and the "test set." The training consists of those samples that we use in our calculations with Bayes' rule. Once we have the posterior probability, we can apply it to the test samples. Note: the test samples were never used in the calculation of Bayes' rule. In the notebook, we assigned, randomly, 80% of the samples to the training set and 20% to the test set. Why did we choose this ratio? We don't want few training samples because then we remove too many data to build our prediction model. At the same time, we want enough test samples to make the test meaningful.

Figure 3.10 provides a summary of the results of applying the classifier on the test set in a table termed the "confusion matrix." This table contains the scores of the frequencies of how many times the classifier was correct or incorrect in determining the two rock types.

The confusion matrix, however, does not tell the whole story on the accuracy of the classifier. To understand this, let's revisit how this classification was done: we took intrusive rocks as winners when the posterior probability of intrusive rocks is higher than 0.5. Imagine now taking a different threshold. If we take a threshold much less than 0.5, then we classify most locations into intrusive rocks. This means we will likely have no false negatives (no missing intrusive rocks). On the other hand, if we make the threshold very high, we are likely to avoid any false positives (mis-identifying intrusive rocks). What this reasoning shows is that there is a trade-off between false positives and false negatives. We need to address this trade-off. To do so, we will need to come up with two metrics of classification performance.

To see how, let's ask the following two questions:

- What proportion of positive labeling was actually correct?
- What proportion of actual positives were identified correctly?

Let's elaborate on the first question. How precise was the correct labeling? Positive labeling means to label a location with a certain rock type. Negative labeling is labeling the absence of that rock type, but not stating what else it is. To quantify the proportion of correct positive labeling, we define the following metric:

$$\text{Precision} = \frac{\#TP}{\#TP + \#FP} \tag{3.19}$$

where TP means, true positive. In this metric, we do not focus on negative labeling.

Let's move to the second question. The actual positives consist of two elements: the true positives and the false negatives (mis-identified positives). Now, we do involve the negatives, which was not the case in precision, and define the concept of recall:

$$\text{Recall} = \frac{\#TP}{\#TP + \#FN} \tag{3.20}$$

Why do we focus on these two? Why not just worry about being "precise," which sounds good enough? To understand the need for both, we revisit the Cape Smith case. Recall that we have divided the dataset into a training set and a test set. We focus on the samples in the test set. For these samples, Bayes' rule provided us with a probability of an intrusive rock (and hence

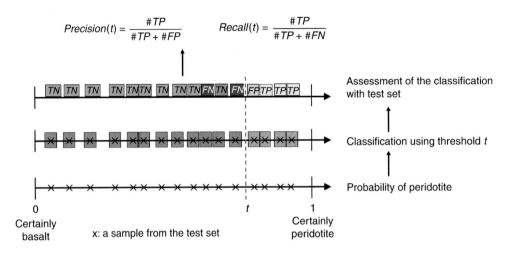

$$Precision(t) = \frac{\#TP}{\#TP + \#FP} \qquad Recall(t) = \frac{\#TP}{\#TP + \#FN}$$

Figure 3.11 We start at the bottom of the graph and map the probabilities of there being intrusive rocks in the test set on an axis. Then we select a threshold t, and call everything below the threshold basalt, and every sample above the threshold peridotite. Then we check the classification result with the actual label in the test set and assign TN, FN, FP, and TP; then we calculate precision and recall.

basalt). We can rank these probabilities from small to large. This is illustrated conceptually in Figure 3.11.

For those samples that are close to zero, we are very certain that we don't have intrusive rocks, so we won't have any false negatives. For samples with very high probability, we are likely to call those correctly intrusive, and hence have very few false positives. Consider now putting a threshold on the probability, and if the probability is less than that threshold, we call it intrusive rock, otherwise, we call it basalt. However, we also know the true outcome; hence, we know whether we created a false positive, true positive, or true negative. For a very high threshold, we will have very high precision, but low recall, while for a very low threshold we have the opposite; hence, the trade-off between precision and recall.

Now we vary the threshold from small to large: for each instance of the threshold, we will get different FP, TP, and FN, and hence we can calculate the precision and recall as function of the threshold. Then, we plot the combinations of precision and recall into a single plot, as shown in Figure 3.12. In this plot, we typically see precision decrease as recall increases, and vice-versa, reflecting the trade-off between these two. The perfect predictor is one that has high precision for all recall values. This plot can be used to compare different classifiers. The closer the curve to the perfect predictor, the better the classifier.

  Visit **Notebook 14: Bayes** to see how Bayesian classifiers are evaluated.

Protocol for Bayesian Prediction and Classification with a Continuous Predictor

Previously, we worked with a binary variable (intrusive rock versus basalt). The same method applies to any amount of class labels. The difference is the following:

Box 3.1 The protocol for Bayesian prediction and classification

- Perform kernel density smoothing of all histograms to get likelihood distributions
- Turn all categories into binary indicator variables
- Calculate or state priors of each indicator
- Calculate total probabilities for each indicator
- Calculate posterior distributions for each indicator
- Plot posterior distributions on a map
- Perform Bayes' classification and create a map of classified rocks
- Create a Bayesian confusion matrix
- Create a precision versus recall curve

Plots to make

- As many histograms of the predictor as categories
- As many posterior probability maps as categories
- As many classified maps as categories
- Bayesian confusion matrix
- Precision versus recall curve

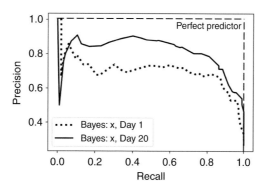

Figure 3.12 Precision versus recall on a test set for Cape Smith with a single predictor. A perfect predictor would return the dashed line.

- Make sure you define indicators of categories and not just work with categories
- Make sure you make spatial maps of posterior probabilities and classifications. You will make as many maps as you have categories. Maps create spatial context.

Box 3.1 summarizes the protocol for Bayesian prediction and classification.

▶ Play **Video 06: Bayes** to learn about Bayes' rule and Bayesian reasoning.

3.4.2 Naïve Bayes' Classification

In many situations, for example as we outlined for landslides and mineral resources, we have multiple evidence, i.e., multiple datasets, or multiple predictors. This multiple evidence problem now becomes much more challenging.

Multiple Evidence

To understand this, consider the extension of Bayes' rule that involves two types of evidence (data):

$$P(H|E_1, E_2) = \frac{P(E_1, E_2|H) \times P(H)}{P(E_1, E_2)} \tag{3.21}$$

We still have the same prior, but now we need to know $P(E_1, E_2|H)$: the likelihood of observing both types of evidence together, given the hypothesis H is true. Previously, we framed evidence as a test or experiment, for example some device to detect a disease or a deposit. Now we have two such tests/devices, and we need to understand how both of them behave together under a changing hypothesis. For example, imagine two ways of measuring or detecting an orebody, by measuring a magnetic signal and a gravity signal. Now we need to know how both behave jointly, not just individually, when there is a deposit and when there isn't. You can see that this gets even more complicated when we also observe many other "signals," a geochemical signal, a geological signal, etc.

The problem is challenging when we have data of very different nature, for example one piece of evidence is geophysical, and another is mineralogical. You may have geophysical databases and mineralogical databases but no cross-data bases, or, if you have, the information is too sparse to find the dependencies you are looking for. What we present below is a way out of this situation; but, of course, this will require making assumptions, and the problem is that these assumptions are difficult to check.

Conditional Independence

What Is Conditional Independence?   The first and most important aspect of conditional independence is that it describes the relationship between, at minimum, three variables. Conditional independence, and this is a source of confusion, is very different from independence, which only needs to involve two variables. Independence is simply as follows: two variables or events are completely independent, and there is no relationship of any kind between them. The cooking skills of Americans on a scale from 1–10 have nothing to do with the amount of people that vote in the next US election. This makes logical sense. In probability, this is expressed as follows:

> P(a random person has cooking skill 8 *and* 110 million people vote in the election)
> $= P$(a random person has cooking skill 8)
> $\times P$(110 million people vote in the election)

Mathematically, it means that you can multiply the probability of these events.

$$P(\text{event 1 } and \text{ event 2}) = P(\text{event 1}) \times P(\text{event 2})$$

In conditional independence, there is "third party." It is therefore incorrect (incomplete) to state: event 1 and event 2 are conditionally independent. So, who is this third party?

A very simple and intuitive example is to look at temperature prediction at Stanford University in the summer. Stanford, being in a semi-arid region with no summer rain, has a fairly predictable temperature; if you know the temperature today, a very good guess is to take that value as a prediction for the temperature tomorrow (those who have seen the movie *LA Story* know exactly what we are talking about!). We can therefore say that "event 1 = temperature tomorrow" is very related to (dependent on) "event 2 = temperature today," for summer days. Now consider the temperature yesterday, our third-party event: "event 3 = temperature yesterday." Because the weather is so constant, we really don't need to know the temperature yesterday to predict the temperature tomorrow; knowing the temperature today is enough:

$$P(\text{event 1} \mid \text{event 2, event 3}) = P(\text{event 1} \mid \text{event 2})$$

It looks now that event 3 and event 1 are not related, but that's absolutely not the case: event 3 and event 1 are in fact very correlated. However, if we had instead an "event 3 = clouds are arriving," we could not make the statement of conditional independence, because clouds arriving is different from temperature measurements; it bring more information to the table, which we cannot ignore:

$$P(\text{temperature tomorrow} \mid \text{temperature today, clouds arriving}) \neq$$

$$P(\text{temperature tomorrow} \mid \text{temperature today})$$

There is a second version of this assumption, which can be derived from the first version:

$$P(\text{event 1 } and \text{ event 2} \mid \text{event 3}) = P(\text{event 1} \mid \text{event 3}) \times P(\text{event 2} \mid \text{event 3})$$

This is less intuitive perhaps, but equally important. Knowledge of "raining yesterday" is enough to predict "raining today" and "raining tomorrow" separately. You don't need to do it in one go, you can separate the problem into two problems, and these problems will be easier to solve. If conditional independence holds true (big if!), we can easily do a calculation with Bayes' rule when we have multiple datasets. This is termed a Naïve Bayes' classifier (under the dangerously naïve conditional independence assumption). So before doing that, we need some way to evaluate this rather strong assumption.

Evaluating Conditional Independence   We now issue a warning! Conditional independence is a very convenient assumption which makes calculations very easy; but at the same time it is very challenging to evaluate whether it applies to your case, which makes it a dangerous assumption as well. Why is that? In order to evaluate if conditional independence applies, we need to compare it with the case where we do not make that assumption:

$$P(\text{event 1 } and \text{ event 2} \mid \text{event 3}) \text{ with } P(\text{event 1} \mid \text{event 3}) \times P(\text{event 2} \mid \text{event 3})$$

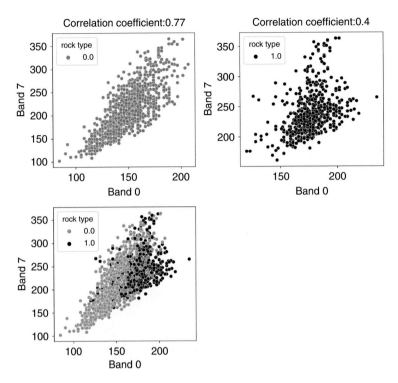

Figure 3.13 Correlation between two bands in remote sensing and how the relation changes when looking at different rock types.

If we can estimate the joint P(event 1 *and* event 2 | event 3), then we would just use it, and forget about making the conditional independence assumption altogether. The test is still useful when you are dealing with more than two types of evidence/predictors. In that case, we can look at how well conditional dependence applies when looking at two bits of evidence at a time.

For now, let's say we have only two types of evidence or data x_1 (band 0, red) and x_2 (band 7, near infrared), as shown in Figure 3.6. Prior to any prediction we can make a scatter plot of x_1 versus x_2 for both intrusive rocks and basalt (Figure 3.13). We can calculate the correlation coefficient and find it to be low (0.4). Now we look at only basalt. In doing so, we created a condition: conditional on the fact that we are only looking at basalt, what is now the correlation? The correlation has changed! Hence the variation between x_1 and x_2 depends on a third party.

Visit **Notebook 14: Bayes** to study pairwise conditional independence in the Cape Smith dataset.

Why does testing of conditional independence matter? It is important for predicting. We will first make a prediction not assuming any conditional independence. We run our Bayesian protocol:

- Prior $P(H_i = 1)$
- Likelihood $f(x_1, x_2 | H_i = 1)$

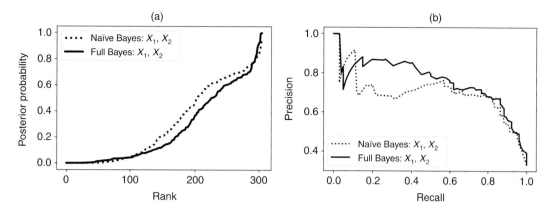

Figure 3.14 (a) Ranking the testing set sample according to the difference in posterior probability between Naïve and full Bayes' classifications. The probabilities are ranked from small to large. (b) Full Bayes' classification yields a better precision–recall correlation relative to Naïve Bayes' classification.

- Total $f(x_1, x_2) = f(x_1, x_2 | H_1 = 1) P(H_1 = 1) + f(x_1, x_2 | H_1 = 0) P(H_1 = 0)$
- Posterior:

$$P(H_1 = 1 | x_1, x_2) = \frac{f(x_1, x_2 | H_1 = 1) P(H_1 = 1)}{f(x_1, x_2 | H_1 = 1) P(H_1 = 1) + f(x_1, x_2 | H_1 = 0) P(H_1 = 0)} \qquad (3.22)$$

The term $f(x_1, x_2 | H_1 = 1)$ is modeled using kernel density estimation.

We can also repeat the same Bayes' protocol to perform a Naïve Bayes' classification, now adding the conditional independence assumption, meaning we replace $f(x_1, x_2 | H_i = 1)$ with $f(x_1 | H_i = 1) f(x_2 | H_i = 1)$. Once we have a probability model $P(H_1 = 1 | x_1, x_2)$, we can perform the classification and calculate the precision versus recall curve. Figure 3.14 shows that the curve for the full Bayes' classification (not assuming conditional independence) lies above that using the Naïve Bayes' method (assuming conditional independence), hence the former is a better classifier. In fact, we notice that the Naïve Bayes' classifier with two types of evidence is not better than when there is just one piece if evidence (Figure 3.12).

Naïve Bayes' Formulation

The Naïve Bayes' formulation involves applying Bayes' rule but assuming conditional independence. In that way, we simply extend the idea of the previous section from considering two pieces of evidence to multiple types of evidence:

$$f(x_1, x_2, \ldots, x_N | H_i = 1) = f(x_1 | H_i = 1) f(x_2 | H_i = 1) \ldots f(x_N | H_i = 1) \qquad (3.23)$$

This means we need to model the likelihood function $f(x_n | H_i = 1)$ as many times as we have evidence.

Visit **Notebook 14: Bayes** to see the predictions for intrusive rocks using all eight bands of remote sensing.

Sensitivity Analysis with Bayes

In addition to predicting, it is also important to know what mattered most in the predictions. It is likely that not all evidence layers are equally important, but how can we quantify their importance?

Let's think about what information does. It changes probabilities! Hence, the more probabilities are changed, the more important that information is. If there is no change, then that information is completely useless. We can know this change using Bayes' rule. Let's start with having no evidence, $P(H = 1)$ and consider a single source of evidence x_1 (e.g., band 0): $P(H = 1|x_1)$. What we wish to do is quantify the difference. The best way to do this is using the logit function:

$$\text{MAD} = \frac{1}{N} \sum_{\text{all samples}} \left| \text{logit}(H = 1|x_1) - \text{logit}(H = 1) \right| \tag{3.24}$$

where MAD stands for mean absolute deviation. We take the absolute value, since we are interested in any deviation from the prior. Going lower on the logit means going closer to $H = 0$. Figure 3.15a shows this MAD for three bands. Since a MAD is a logit, we can turn it back into a probability; MAD $= 0$ is the same as $p = 0.5$, so in this figure we look at values exceeding 0.5.

We can use this concept to look at any combination of evidence layers. For example, we can compare the Bayesian prediction for all the evidence with the prior:

$$\frac{1}{N} \sum_{\text{all samples}} \left| \text{logit}(H = 1|x_1, x_2, x_3, \dots) - \text{logit}(H = 1) \right| \tag{3.25}$$

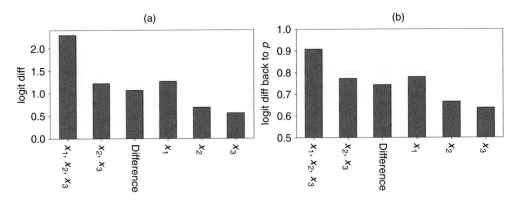

Figure 3.15 Mean absolute deviation in (a) logit terms and (b) transformed back to the probability.

Additionally, we can assess how well adding one bit of evidence does when added to the dataset. Imagine that we only have three bands (band 1, band 2, and band 3) as evidence layers, and we want to know how much information band 3 gives us. To do that, we calculate the following MAD:

$$\frac{1}{N} \sum_{\text{all samples}} \left| \text{logit}(H = 1|x_1, x_2, x_3) - \text{logit}(H = 1|x_1, x_2) \right| \tag{3.26}$$

Figure 3.15 show various MAD values. Band 1 is more informative than band 2 and band 3, as shown in the posterior higher probability and logits. Note in the figure the MAD values are calculated for various combinations of the evidence (band data): the "difference" refers to:

$$\left| \text{logit}(H = 1|x_1, x_2, x_3) - \text{logit}(H = 1|x_1, x_2) \right|. \tag{3.27}$$

Protocol for Naïve Bayes' Classification

Box 3.2 gives the protocol for performing a Naïve Bayes' classification. Note that we are dealing with a multivariate dataset as predictors; in our case, these are the multiple bands. For any

---

Box 3.2 The protocol for Naïve Bayes' classification

- Perform **the protocol for principal component analysis (PCA)** on the predictors to get a better insight into the data
- Identify any multivariate outliers in the predictors, talk to expert about what to do with them
- Perform a two-way conditional independence analysis
- Split the data into a training set and test set
- Calculate prior probabilities based on the label data
- Smooth histograms of predictors per each class using kernel smoothing
- Calculate total probabilities
- Calculate posterior probabilities for each class using Bayes' rule under the assumption of conditional independence
- Perform the classification: take the highest probability as the winning class
- Assess the quality of the classification using a Bayesian confusion matrix and a precision versus recall analysis
- Map the probabilities and the classified labels into two-dimensional maps

Plots to make
- Any plots related to multivariate analysis
- Two-way conditional independent analysis
- Histograms of predictors for each label
- Bayesian confusion matrix and precision versus recall plot
- Spatial maps with probabilities and classification results (one map for each class)

multivariate or also a compositional dataset, we need to execute the protocol of multivariate or compositional analysis a priori.

3.4.3 What Have We Learned in Section 3.4?

- We have learned about Bayes' rule, a simple formula that can predict how the probability of something of interest changes when new information is acquired.
- We have learned about the counterintuitive nature of Bayes' rule: how Bayes' rule shows the surprising difference between the power of a confirming and debunking test to assess a leading hypothesis or theory.
- We have learned about the important concept of conditional independence: an assumption that is made concerning the relationship between three different events or variables.
- We have learned about ways to assess whether this assumption applies in a dataset.
- We have learned about ways of comparing various predictions using a precision and recall analysis.

3.5   Linear and Logistic Regression

Bayes' rule provides us with a general framework for prediction and uncertainty quantification. This rule acknowledges that, in most prediction problems, there are two sources of information: what we already know about something of interest before getting the data (prior) and getting more data to improve the prediction (likelihood). The way knowledge is quantified is through probabilities (for discrete variables) or probability densities (for continuous variables). A disadvantage of Bayes' rule is that it becomes less easy to apply when we have a large number of variables. We acknowledged that issue when we assumed conditional independence; this is a convenient but hard-to-test assumption, but it is needed to apply Bayes' rule easily when there are a lot of variables.

In this section, we look at an alternative approach. To understand how this works, let's focus on what we eventually want to get out of all this. Ultimately, geologists want a map with probabilities of rock types which could guide them in visiting highly probable areas of rocks of interest (intrusive rocks). What they need is the posterior probability of a rock type. Bayes' rule offers a way to get this posterior probability by quantifying the prior and the likelihood. Here we offer a new method: estimating the posterior probability directly using regression methods.

3.5.1 What Is Regression?

Much of data science is about finding relationships between variables, then using these relationships to make predictions with associated uncertainty quantifications. The concept of "regression" is an important tool for doing this. The word itself seems quaint and its meaning perhaps isn't intuitive. In non-technical terms, it means "the act of regressing," or "returning to a lower or less perfect state."

The way this word made it into the statistical world is rather serendipitous. We need to go back to 1886 and the British anthropologist Francis Galton, who published a paper entitled "Regression towards mediocrity in hereditary stature" (Galton, 1886). Galton was cousin of Charles Darwin and was in fact the mentor of Karl Pearson, the well-known English mathematician and biostatistician. In those days, many statisticians were working in the field of eugenics, unfortunately. Fisher himself, often considered as one of the founders of statistics, said: "human groups differ profoundly 'in their innate capacity for intellectual and emotional development'" (Evans, 2020). The first paper on regression can therefore be framed in this historic context. Unlike Fischer, however, Galton had not developed any rigorous statistical method. His paper describes a study comparing the heights of children with the heights of their parents, to gain insight into what degree height is an inherited characteristic.

Let's look at Galton's study. The data are shown in Figure 3.16, defined by bins of parents' heights and the calculated mean child's height for each bin. (Note that when he published the data, Galton literally eye-balled – there was no fancy software then – a straight line.) What Galton was interested in was the slope, namely whether it was larger than, equal to, or less than 1? This would, in his view, indicate a change between generations.

A slope of less than one indicates regression (now in the sense of "less perfect state"), with children having more mediocre heights than their parents. Unfortunately, Galton forgot a simple check: what if we make X = children and Y = parents. If there were symmetry, then fitting the line that way would result in a slope larger than 1; it wasn't (see the slope values in Figure 3.16b).

The name "regression" (now referring to "regression to the mean") was born. In such regression, there is no symmetry in the Y and X variables. The slope of Y over X is not the reciprocal of the slope of X over Y; the reason is simple: when you fit $Y|X$ you can see the variation in Y relative to a straight line (the error in Y); when you fit $X|Y$, you see the error in X.

Regression therefore requires that you state what you are fitting and how you are fitting it. In statistics and machine learning the latter – "how you are fitting it" – is termed "loss function." The loss is not referring to losing something. It refers to the remaining variation that exists when

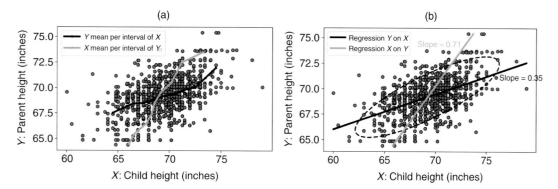

Figure 3.16 Galton's regression data in modern software, showing (a) the conditional mean of $Y|X$, and of $X|Y$; and (b) the linear regression. A black and white version of this figure will appear in some formats. For the color version, refer to the plate section.

fitting the line. That variation can be measured in values of Y; this would give you a regression of Y on X. We call X the "predictor" (that allows us to make predictions) and Y the "predictand" (that is being predicted). Sometimes, researchers use the term "independent variable" for X. We don't really like using this term here: it suggests an independence that may not be there. This is something we will return to when we have not just one type of X variable (height) but multiple types of X (height, weight, age, etc.).

> Visit **Notebook 15: Linear Regression on Galton Data** to study what data Galton used and to study the asymmetry in regression.

3.5.2   Linear Least-Squares Regression

In regression we have data on X and Y as pairs of observations $(x_1, y_1), (x_2, y_2), \ldots, (x_n, y_n)$. When fitting a linear function to a set of data, we often first state a model that we believe explains the relationship between X and Y:

$$Y = \beta_0 + \beta_1 X + \varepsilon \tag{3.28}$$

In other words, we assume the relationship between X and Y consists of a deterministic part with unknown coefficients β_1 and β_0 and a random part ε. It is important to recognize that this is a model, and the data at hand may not follow this hypothesized reality; hence, after fitting the data, we need to check that indeed the errors are completely random.

Fitting the Linear Function

To fit a linear function, we use the concept of (ordinary) least squares; it simply means that we define the difference between the linear function and the data as

$$\text{Loss}(\beta_0, \beta_1) = \sum_{i=1}^{n} (y_i - \beta_0 - \beta_1 x_i)^2 \tag{3.29}$$

This function, also termed the objective function or loss function, is a function of β_1 and β_0, as shown in Figure 3.17. We notice in this figure that there is a minimum. Analytical expression can be derived to calculate this minimum; in the case of Figure 3.17 we find:

$$(\hat{\beta}_0, \hat{\beta}_1) = (45.2067; 0.3467)$$

Here, the hat on the β indicates that this is an estimate.

In case of multiple X values, we extend the model to:

$$Y = \beta_0 + \beta_1 X_1 + \beta_2 X_2 + \beta_3 X_3 \ldots + \varepsilon \tag{3.30}$$

and estimate the β values by now minimizing:

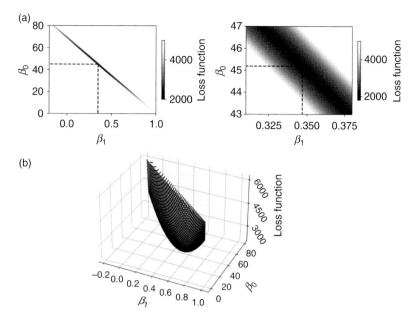

Figure 3.17 (a) Two-dimensional and (b) three-dimensional views of the loss function of the linear fit as a function of β_1 and β_0.

$$\text{Loss}(\beta_0, \beta_1, \beta_2, \beta_3, \dots) = \sum_{i=1}^{n} \left(y_i - \beta_0 - \beta_1 x_{1,i} - \beta_2 x_{2,i} - \beta_3 x_{3,i} - \dots \right)^2 \quad (3.31)$$

Assessment of Residuals

We return to the fact that, in fitting a linear function to the data, an assumption was made that, after fitting, the error is random. How do we check this? We do so by a diagnostic tool termed analysis of the residuals, or residual analysis. Figure 3.18 shows such an analysis for the linear regression of the Galton dataset. Ideally, and that is the case in Figure 3.18, we want the residuals to be evenly distributed along the x-axis. What we don't want, as shown in Figure 3.19, is:

- a change in variation over the x-axis
- a non-linear variation
- the existence of outliers

Any of these three will violate the random error assumption.

Expressing Confidence in Linear Regression

If we cleared the hurdle of the residual analysis, we next apply some additional tools to assess the fit. The first assessment concerns the uncertainty on the coefficients β. Clearly, with fewer

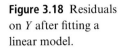

Figure 3.18 Residuals on Y after fitting a linear model.

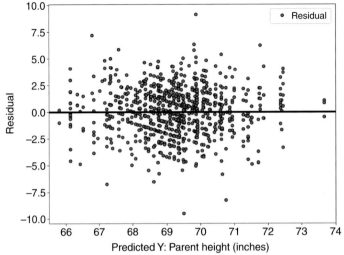

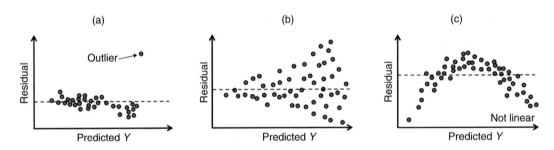

Figure 3.19 Three hypothetical examples where the residuals of a linear regression indicate the model is not followed: (a) the existence of outliers; (b) the variance of the residual changes with x; (c) the residuals exhibit non-linear behavior.

data, these coefficients will be less certain (more variation allowed). A bootstrap analysis (see Section 5.2.3) provides us with a range of estimates of β coefficients; see Figure 3.20. As shown, the estimates of the coefficients are correlated. Indeed, when making a slight change of one β coefficient, we can make another small change to another β coefficient and still get a reasonable fit to the data. This correlation between the estimates of coefficients in linear regression is very common.

In Figure 3.20 we are particularly interested in the histogram of β_1. In Eq. (3.30), this coefficient sits in front of the prediction variable X; hence, if it is significantly larger than zero, X is a good predictor of Y. This is clearly the case in the Galton dataset.

A second way of summarizing the uncertainty of the β coefficients is shown in Figure 3.21. The black line is the regression line, while the gray area tells us where the regression line falls when varying β (here between the 2.5 and 97.5 quantiles, as shown in Figure 3.21a). On Figure 3.21b, we show the variation of the residual around the best fit. The grey area here is the 95% confidence interval (CI) for the prediction of Y, given any value of X.

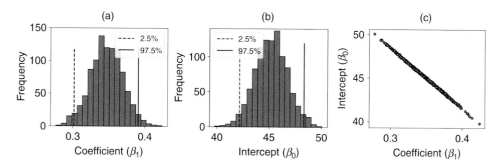

Figure 3.20 Uncertainty on each β coefficient obtained from the bootstrap method: (a) and (b) histograms representing the uncertainty of β_1 and β_0; (c): the correlation between β_1 and β_0.

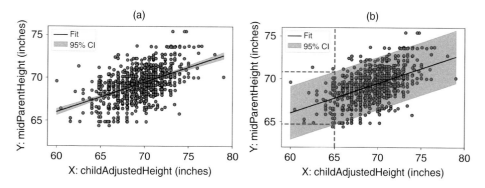

Figure 3.21 (a) The 95% confidence interval (between the 2.5 and 97.5 quantiles) on the regression line. (b) The 95% confidence of the prediction. For example, if $X = 65$, then we can say with 95% confidence that Y is roughly between 64.5 and 71.

 Visit **Notebook 15: Linear Regression on Galton Data** to study linear regression and how we represent uncertainty on linear regression and predictions made using linear regression.

3.5.3 Logistic Regression

Linear regression works when variables are continuous and, of course, a linear relationship is hypothesized. However, our target is not some continuous property, but a probability: the probability of a rock type given the remote sensing data. We cannot use linear regression for this purpose because what we want to predict is discrete: a "0" (basalt) versus a "1" (intrusive rock).

The Logistic Function

Let's briefly return to the Galton dataset. We calculate the mean of y for some interval of x. How does that work for binary variables? What in fact is the mean of a binary variable? It is the probability of the binary variable being one. In other words, the mean is a probability.

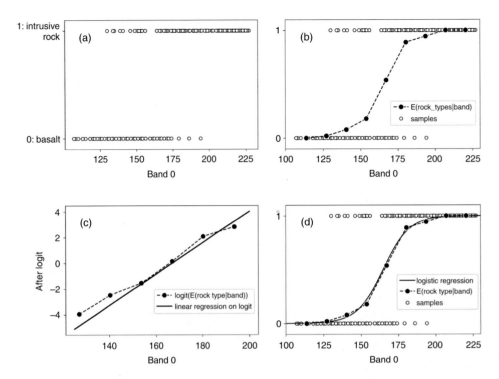

Figure 3.22 (a) Cape Smith data for one band (the values on the *x*-axis are for band 0). (b) The mean calculated by binning the band data. (c) The same Cape Smith data after the logit transformation. (d) Logistic regression with one band returns the probability of intrusive rocks given the band data.

Statisticians would say that the expected value of Y (binary) given a continuous X is the probability that $Y = 1$ given X; mathematically, this is:

$$E[Y|X] = 1 \times P(Y = 1|X) + 0 \times P(Y = 0|X) = P(Y = 1|X) \qquad (3.32)$$

The idea of regression to the mean is still a good one, we just need to make a few modifications. To see how, let's return to the Cape Smith data and again look at only one remote sensing band, as we did above. We plot the X (band 0) on the *x*-axis and the rock type on the *y*-axis (Figure 3.22a). We note that the Y is indeed binary. Now we bin the X and calculate the mean of the binary variable. What do we observe? First the shape that emerges is not linear (a straight line), it curves (Figure 3.22b). How does it curve? To diagnose the curve, we return to log ratios. More specifically, we calculate the log of the odds ratio (the logit):

$$\text{logit}(p) = \log\left(\frac{p}{1-p}\right) \qquad (3.33)$$

This is shown in Figure 3.22c, where we observe a linear relationship between the logit and the remote sensing band. You can also invert it using the logistic function (see Figure 3.22d), as discussed in the section above on comparing probabilities:

Table 3.2 An example of a dataset and two competing logistic models

Data			Model			
Band 1 value	Band 2 value	H	Prediction $P(H)$ model 1: $H = 1$	Prediction $P(H)$ model 1: $H = 0$	Prediction $P(H)$ model 2: $H = 1$	Prediction $P(H)$ model 2: $H = 0$
−1	2	1	0.8699	0.1301	0.3100	0.6900
−1.5	3	1	0.9453	0.0547	0.2315	0.7685
4	−1	0	0.3543	0.6457	0.9802	0.0198
Likelihood = make product			0.5308		0.0014	

$$\text{if } x = \log\left(\frac{p}{1-p}\right), \text{ then } p = \frac{1}{1 + \exp(-x)}$$

Logistic Regression by Maximum Likelihood

Fitting a logistic regression model differs from least-squares fitting in that we can no longer use the concept of random residuals. We will skip any mathematics here and instead focus on this new concept through an illustrative example. Consider a dataset with three samples and two variables, with the aim of predicting a discrete variable (see Table 3.2). Imagine we have two competing logistic models:

$$\text{model 1:} \quad (\beta_1, \beta_2) = (0.1, 1)$$
$$\text{model 2:} \quad (\beta_1, \beta_2) = (1, 0.1)$$

We'd like to quantify which of these two models is a better fit to the data. In other words: which model provides a better prediction of the labels (H). To do so, we use the logistic model:

$$P(H = 1) = \frac{1}{1 + \exp(-(\beta_1 \text{band } 1 + \beta_2 \text{band } 2))} \tag{3.34}$$

We can compute this for all three samples and for both competing models. We notice in this table that model 1 predicts the data better because it has high probabilities on $H = 1$ when the sample is intrusive, and it has high probabilities on $H = 0$ when the sample is basalt. To get to a single metric, we take the product of these probabilities. The higher this product, the better the fit. Model 1 fits better because it puts more weight on the important variable, band 1.

The concept used here is likelihood. The product of probabilities is the likelihood of the data under some hypothesized models. This likelihood is a function of β; therefore, maximizing the likelihood will return the maximum likelihood estimates of β.

Multiple Logistic Regression

Multiple logistic regression is similar to multiple linear regression. But we find β in the logistic model:

$$\log\left(\frac{P(H=1)}{1-P(H=1)}\right) = \beta_0 + \beta_1 x_1 + \beta_2 x_2 + \ldots \tag{3.35}$$

We then compute the maximum likelihood estimates of β that best explain the data. Once the β values are estimated, we can make predictions:

$$P(H=1) = \frac{1}{1 + \exp(-(\beta_0 + \beta_1 x_1 + \beta_2 x_2 + \ldots))} \tag{3.36}$$

Logistic Regression on Cape Smith Data

 Visit **Notebook 16: Logistic Regression** to study the application and evaluation of logistic regression for Cape Smith.

How does logistic regression compare with the Naïve Bayes' method? To check this, we plot the precision versus recall curve when using all bands, for day 1 and day 20; see Figure 3.23. In this case, both methods perform equally well. However, what is the difference with Naïve Bayes classification? Let's make a methodological comparison and remind ourselves of the assumptions being made in either case.

Comparison with Bayes' Classification

At this point, logistic regression may seem a "better" or "easier" method for prediction than using Bayes' rule. After all, we just feed the data into a logistic regression software package and call it a day. However, logistic regression is not without assumptions.

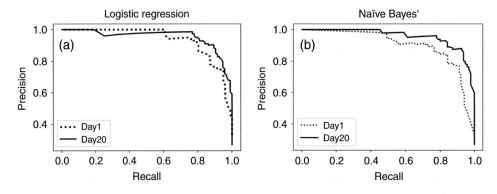

Figure 3.23 Comparison of precision versus recall plots calculated using (a) logistic regression and (b) Naïve Bayes' methods.

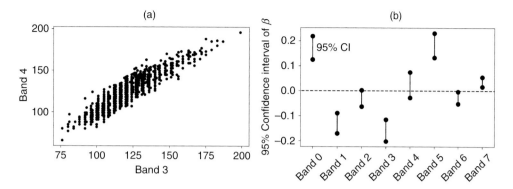

Figure 3.24 (a) Scatter plot between predictor band 3 and band 4 data showing the collinearity. (b) The 95% CI of logistic regression coefficient β for all the predictors.

The advantages of logistic regression over the Bayes' method are the following:

- It is easy to do.
- There is no need for conditional independence assumptions if we use Naïve Bayes' classification.
- It works well for a large number of predictors X, even with a limited amount of data (a rule of thumb is to use a sample size = $100 + 50 \times$ number of variables in X).

The estimated coefficients β can also be interpreted in terms of how sensitive (important) a predictor X is in the prediction of Y (the presence or absence). However, this interpretation only rigorously works when the X values have no dependency, then the β coefficients lose their meaning. To illustrate this, let's examine the effect of dependency in the Cape Smith case. In Figure 3.24a, we observe a strong linear correlation between remote sensing bands 3 and 4. It means they should contribute similarly to the prediction and thus share similar coefficients in logistic regression. But the CI in Figure 3.24b shows this is not the case. They show very different coefficients with opposite +/− signs. The CI of band 4 even includes 0, which means it lacks statistical significance. In the figure, you also notice that the 95% CIs of band 2 and band 6 are very close to 0. When visiting the notebook, you can examine the multi-collinearity of those variables following the provided examples for bands 3 and 4.

The disadvantages of logistic regression compared to the Bayes' method are as follows:

- In Bayes we do not assume a linear relationship between the logit and the predictors X.
- The full Bayes performs better if you can do it, which is usually the case with few predictors. Or, if you use Naïve Bayes' classification, and conditional independence assumption applies well to your dataset.
- Bayes' classification can be done for continuous variables and hence you are not limited to binary variables, as is the case with logistic regression
- In Bayes' classification, correlation between the predictors (X) is accounted for.

Box 3.3 Logistic regression protocol

- Perform the protocol for PCA on the predictors to get better insight into the data
- Identify any multivariate outliers in the predictors, talk to an expert about what to do with them
- Assess any multi-collinearity in data; PCA will help in this analysis
- Split the data into a training set and test set
- Run logistic regression software to output posterior probabilities
- Perform classification: take the highest probability as the winning class
- Assess the quality of the classification using a Bayesian confusion matrix and precision versus recall plot
- Map the probabilities and classified labels into two-dimensional maps

Plots to make

- Any plots related to multivariate analysis
- Identification of correlation between predictors
- Bayesian confusion matrix and precision versus recall plot
- Spatial maps with probabilities and classification results (one map for each class)

Logistic Regression Protocol

Box 3.3 summarizes the logistic regression protocol for spatial data aggregation.

 Play **Video 07: Regression** to learn about regression methods and logisitic regressions.

3.6   What Have We Learned in This Chapter?

- We have learned that predicting rare events is difficult even in the presence of very accurate information.
- We have learned to think "Bayesian," a general way of thinking about predictions/uncertainty that involve information.
- We have learned about two competing approaches to make prediction using information: the Bayesian approach versus the regression approach.
- We have learned that regression approaches, while possibly less accurate than a full Bayesian approach, can also provide good predictions, and are easier to apply.

Both approaches are foundations to more advanced techniques in the area of machine learning. For example, neural network extensions are available for logistic regression, where the logit model is no longer assumed linear. Bayesian approaches are extended through a mixture of density networks. Here, the simple kernel density smoothing is replaced by more sophisticated density models that operate well in dimensions higher than 10. Regardless of these more advanced techniques, it is always useful to start with these simpler methods and build from there!

TEST YOUR KNOWLEDGE

3.1 Why are there so many false positives in resource exploration?
 a. Because many other geological phenomena act like resources
 b. Because geophysical data are very inaccurate
 c. Because of lack of drilling
3.2 What would you worry most about in predicting hazards?
 a. False positives
 b. False negatives
 c. True positives
 d. True negatives
3.3 What is prior information?
 a. What you know about something before acquiring new information
 b. How probable the new information is before acquiring it
 c. Knowing what the leading hypothesis is before testing
3.4 Where do we get prior information on a hypothesis?
 a. By calculating the probability of the evidence
 b. By asking experts
 c. By means of high-performance computing
3.5 Why should you switch doors in the Monte Hall problem?
 a. Because the show host knows where the car is
 b. Because the contestant knows where the car is
 c. Because they both know where the car is
3.6 What is the difference between independence and conditional independence?
 a. Conditional independence involves at least three variables or events
 b. Conditional independence is a stronger assumption
 c. There is only a difference in the binary case, it does not apply to the continuous case
3.7 Which method provides a more accurate sensitivity analysis?
 a. Bayes'
 b. Logistic regression
 c. Both are equal
3.8 What is the most important reason to use logistic regression over a full Bayes' classification?
 a. It is easy to use
 b. No conditional independence is needed
 c. It works for correlated samples
3.9 When should you use Naïve Bayes' instead of full Bayes' classification?
 a. When you have more than 10 variables

      b. When you are predicting a rare event

      c. With multi-collinearity

3.10 Why do we use the logit function to calculate probabilities?

      a. Because probabilities are like a composition, they sum to 1

      b. To make the product of two probabilities

      c. To be able to apply Bayes' rule

FURTHER READING

We recommend two textbooks for the further readings on Bayes' classification:

- Sivia, D. and Skilling, J. (2006). *Data Analysis: A Bayesian Tutorial* (2nd edition). Oxford University Press.
- Stone, J. V. (2013). *Bayes' Rule: A Tutorial Introduction to Bayesian Analysis.* Sebtel Press.

Stone (2013) provides many interesting examples in the first chapter.
For logistic regression, the following textbook provides a comprehensive guide:

- Hosmer Jr, D.W., Lemeshow, S., and Sturdivant, R.X. (2013). *Applied Logistic Regression.* John Wiley & Sons.

We also recommend the following journal article, which reviews logistic regression and beyond.

- Dreiseitl, S. and Ohno-Machado, L. (2002). Logistic regression and artificial neural network classification models: a methodology review. *Journal of Biomedical Informatics*, 35(5–6), 352–359.

REFERENCES

Evans, R. J (2020). RA Fisher and the science of hatred. *New Statesman*, July 28. www.newstatesman.com /uncategorized/2020/07/ra-fisher-and-science-hatred

Galton, F. (1886). Regression towards mediocrity in hereditary stature. *The Journal of the Anthropological Institute of Great Britain and Ireland*, 15, 246–263.

Lawley, C. J., Tschirhart, V., Smith, J. W., et al. (2021). Prospectivity modelling of Canadian magmatic Ni ($\pm$Cu$\pm$Co$\pm$PGE) sulphide mineral systems. *Ore Geology Reviews*, 132, 103985.

4 Geostatistics

Expected Learning Outcomes

- You will learn about a comprehensive analysis for spatio-temporal data that does not require data to be on a regular grid.
- You will learn about the variogram, a spatial analysis tool that allows us to summarize complex spatial variation with very few parameters.
- You will learn how such variograms can be used to perform spatial interpolations that do not show any artifacts near data and make use of spatial variation in datasets.
- You will learn how any method of spatial interpolation creates models that have less variance than the original dataset and why this is a problem in real-world applications.
- You will learn about methods of conditional simulation that create many maps and that exhibit spatial variability of the dataset.
- You will learn to apply this to relevant earth surface and subsurface questions: the melting of Antarctica and enhancing sustainable farming in Denmark.

4.1 Introduction

The study of geoscientific processes and analysis of geoscientific data takes place in space and time. Geostatistics is a part of statistical science that provides tools for spatio-temporal data analysis and predicting in space–time. The random variables of previous chapters are now a function of space and/or time. In this book, we will limit ourselves to three-dimensional space only, knowing, however, that the same approach works by adding the time dimension or, in fact, in any dimension larger than one.

Let's ask ourselves this question: What is so special about space? The fabric of space is infinitely large! We can take any area and keep subdividing it into ever smaller pieces. This means that we must find clever ways to handle this concept when developing and using tools for spatial data analysis. We may want to study data without the need to discretize data onto a grid. The coordinates of a particular data value, such as the elevation at some location x, y, are real values. While in many applications the final product of a spatial analysis is some grid, the underlying mathematical treatment should not start from any grid.

Figure 4.1 The variance of the logarithm of gold grade values as a function of the size of the area measured (from Krige, 1952).

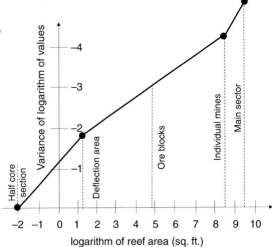

The second important concept in space is that of the "volume support" of the data. Each sample we collect, whether porosity, water quality indices, or copper grade, is taken over a finite volume. This volume can be very specific and direct; for example, a core, or a soil sample. However, volumes may be indirectly specified. Geophysical imaging is one such example. Geophysical signals, when transmitted into the Earth from the surface, weaken (attenuate) as they go deeper. This means that they can image less and less clearly a property of interest. The volume support here refers to the smallest scale at which the geophysical image can detect something; it increases with depth and the measured resolution decreases with depth. The birth of the idea of geostatistics started with a mining engineer, Danie Krige, who studied variances of gold grades when averaged over increasingly larger areas, or larger volume supports, see Figure 4.1. The relationship between the variance of a variable and the domain over which it is averaged is a function of the spatial variability in that area. This concept is important in mining, where large mining blocks are evaluated by means of drillholes with much smaller volume support.

Geostatistics therefore has its roots in the mining industry. Krige (a mining engineer) and his colleague H. S. Sichel (a statistician) found that when applying traditional statistical methods to gold deposits, erroneous results occurred. These traditional methods do not take account of the nature of how gold is distributed (often along veins) in the subsurface. Their goal was to estimate the distribution of grades on large mining blocks from drillhole information. Here, two effects take place: (1) the blocks have a larger volume than the drill holes, and (2) grades are not randomly distributed. However, it was only later, in 1969 and 1971, that the French engineer Georges Matheron formalized Krige's observations and concepts into a single theory: the theory of regionalized variables. "Kriging" is the name Matheron gave to estimating unknown values at spatial locations from nearby sample data. Kriging is one of the most general forms of linear regression, where all (linear) dependencies are accounted for. Today, we see a resurgence of kriging in computer science and machine learning, in the form of Gaussian process regression.

From a niche statistical science in the 1960s and 1970s, geostatistics has evolved to other areas of spatio-temporal analysis and prediction, meaning basically all geosciences. The increased acquisition of massive spatio-temporal data and the vast increase in computational power is at the root of a different focus in geostatistics today. Fundamental societal problems, such as climate change, environmental destruction, and even pandemics, involve spatio-temporal data. Solutions to these problems will require ingesting vast datasets while treating them in a repeatable, coherent, and consistent framework. Spatio-temporal analysis is more challenging than multivariate analysis. While, as we discussed at length in Chapter 3, in multivariate analysis, covariance matrices are fundamental building blocks, geostatistics requires these covariances to be functions of distances measured in space.

A key point we emphasize in this chapter is that any estimation or regression method results in smooth interpolation. The "smoothing" refers to the fact that an interpolated map has less variation than the data values themselves. The interpolated maps we often look at in papers or presentation are not realistic, in the sense that they have less variation than the data they were constructed with. This problem is addressed head-on in a family of spatial modeling tools termed (spatial) stochastic simulation. Instead of one best-guess map, the aim is to build realistic models of the real world, where realism is, at least initially, based on modeling statistical properties of the data.

In the late 1990s, it became clear that just the measured data of the phenomenon at the study area of interest may not be sufficient and that additional prior information is required. This problem was very acute in groundwater and oil/gas reservoirs. Only very few wells are drilled in the subsurface because they are costly. Since most of these wells are vertical, the three-dimensional phenomenon is well sampled in the vertical dimension but poorly sampled in the horizontal dimension. Yet, geologists know a lot about how geology looks horizontally: they can study it in outcrops or create conceptual models. A new question was posed: how can we use analog information in building models in space–time? The analogy to machine learning is that a model should be trained with data from "something else," then applied, or generalized, to the problem at hand. Since geostatistics deals with spatio-temporal data, the training data would need to be of that kind, somewhat exhaustive in space–time.

Summing up, the following concepts have been introduced:

- Space is special: we need somehow to deal with the fact that coordinates in space–time are continuous and don't need to be on a pre-defined grid.
- The volume of the samples matters. Samples are not infinitesimally small, abstract points in space. Different volumes lead to different variances of the value being studied.
- Geostatistics started in the mining world, where the important problem is to accurately estimate the grade of a block to be mined. This estimation is called "kriging" and we will be covering this in detail.
- In many applications, however, we are not interested in just a single estimate of a value, but in representing the real variation in space with models.
- Often, we have done studies before or have analog information. There is a need to be able to use this information together with all of the above ideas.

4.2   Motivating Examples

To further illustrate the concepts introduced above, we present two case studies with spatial data. The first case study uses radar data to map the subglacial topography in Antarctica and thereby understand the melting challenge there under climate change. The second case study maps groundwater redox conditions in Denmark, utilizing both geophysical imaging and borehole measurements, in order to design the targeted nitrogen regulation in Danish farmlands.

4.2.1   Melting in Antarctica

As the twenty-first century unfolds, the markings of climate change have become only too clearly visible. This is certainly true if you live in California. Droughts, fires, smoke, and landslides are now part of people's daily lives. An even greater challenge looms with sea levels projected to significantly increase over the next century depending on how many degrees of warming there is above pre-industrial levels. Geoscientists are therefore focusing their attention on the two largest masses of ice on land: Antarctica and Greenland. The physical behavior of these large ice sheets under increasing temperature is therefore an important area of scientific research. Such research will rely on building physical models of ice-sheet melting and movement as well as collecting vast amounts of data that will be used to build and calibrate models for predicting the sea-level rise. Evidently, these predictions will be subject to uncertainty because of approximations made in models and because of the limited number of measurements we can acquire. In this chapter, we focus on one aspect of the data, namely measuring the bedrock topography below the ice sheet of West Antarctica. Topography is required to predict ice movement and also subglacial hydrology, namely the creation of subglacial lakes and river systems.

To "see" through the ice and observe the topography, geophysicists use a form of imaging involving radar (radio detection and ranging). This system uses radio waves to determine the distance to an object. Think of airports and aircrafts (airborne radar technology was developed during World War II). Glaciers are particularly well suited for radar imaging because there is very little loss of radio frequencies as it penetrates the ice from the surface. For that reason, the bedrock can be detected several kilometres below the ice. Airplanes, often fixed-wing and propeller-driven models, have antennas installed that project the radio waves down through the ice.

The characterization of subglacial topography is particularly important for the Thwaites glacier in the Amundsen Sea Embayment, which is experiencing accelerating ice loss that could destabilize the West Antarctic ice sheet. The problem of course lies in covering a vast area like West Antarctica, which is about 1.5 times that of California. Since data are collected with aircraft, the data along the flightlines are very dense, while they are very sparse between the flightlines, which are often several kilometers apart. For that reason, very large gaps in the data remain, see Figure 4.2.

The problem therefore is to create interpolations of topography from measured topography along the flightlines. An important additional question that needs to be addressed is around

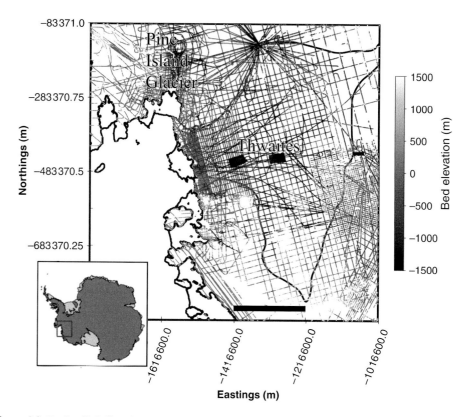

Figure 4.2 Radar flightline data acquired over West Antarctica (from Yin et al, 2022, distributed under a CC BY 4.0 license). A black and white version of this figure will appear in some formats. For the color version, refer to the plate section.

uncertainty. After all, we need to evaluate how this uncertainty affects the uncertainty on sea-level changes. The questions that we want to address are:

- How can we interpolate the flightline data without creating any artifacts that do not exist in nature?
- How can we guarantee that realistic variation is maintained in the interpolation?
- How can we guarantee that we do not get smoother interpolations in areas with few data compared to areas with dense data?
- How can we assess uncertainty on these interpolations?

And most importantly: *How do we propagate this uncertainty to sea-level rise predictions and their associated uncertainty?*

4.2.2 Sustainable Farming in Denmark

Sustainable farming that balances the need for food production with its effect on the environment is a global challenge. An important problem in this context is nitrogen (N) leaching caused by agricultural activities. We might ask why nitrogen is a problem, since it makes up 78% of the Earth's atmosphere! In industrial farming, nitrogen-based synthetic fertilizers are used on a large scale. Several pathways that affect climate and environment are caused by this activity.

The production of nitrogen fertilizers often involves fossil fuels such as natural gas, which is burned to get liquid ammonia. Excess nitrogen causes the production of oxides such as nitrous oxide (Tian et al., 2020), a greenhouse gas 300 times more potent than carbon dioxide. In soils, excess nitrogen causes imbalances in the ecosystem benefitting certain plants at the expense of others. In addition, reactive nitrogen is soluble, it can easily make its way into the groundwater system and eventually in to rivers and lakes, where it causes large algae blooms, creating local "dead zones."

Denmark is a good example of how the effect of excessive nitrogen can be mitigated on a local scale. It has a temperate climate with good soils and plenty of rain. The country has a population of 5.8 million; however, food production is high enough to feed 15 million people.[1] Denmark has 14 million pigs, more than people!

Not all agricultural fields are equally sensitive to the nitrogen problem. Ideally, nitrogen stays local, meaning there is no pathway for it to escape, and the local subsurface has significant capacity to denitrify the soil. How does this work? Nitrogen in fertilizer enters the soil as nitrate (NO_3^-) or ammonium (NH_4^+), where the ammonium is partly taken up by soil or converted into nitrate through nitrification. Denitrification is a process that converts nitrate to nitrogen gas N_2 and hence return is to the atmosphere, in an inert form. Denitrification involves a redox reaction (reduction–oxidation). High denitrification rates are needed in these reactions to make N_2 from NO_3^-. The rates are dependent on what type of minerals and chemical compounds are in the subsurface, more specifically, the presence of denitrifying bacteria that convert nitrate to nitrogen gas. If we know the denitrification potential at different locations or the subsurface redox condition, we can perform spatially differentiated regulation at the local agricultural level, which is a much-needed management strategy. This spatial regulation, often called targeted nitrogen regulation, can meet both the reduction target effectively and mitigate the economic consequence for farmers.

What we want to know is under which fields the subsurface contains a high potential for denitrification. If the subsurface is homogenous (no variations in soil properties), we know three zones exist in sequence: (1) an upper oxic zone with no denitrification potential, (2) a reducing zone with active nitrate reduction, and, finally, (3) a deeper reduced zone with no active nitrate reduction but a potentially high preserved nitrate reduction capacity. This ideal model does not work in heterogenous soils. This is certainly true in Denmark, where the subsurface is dominated by material left by deglaciation in the last ice age. This means that the soil contains spatially varying mineral content, complicated subsurface flows, and a varying presence of bacteria. The goal therefore is to know where the interfaces between the three zones are. In other words, we are looking for surfaces where we see a transition between oxic, reducing, and reduced conditions, and we need to know this at rather small scale (an agricultural field).

---

[1] https://agricultureandfood.dk/

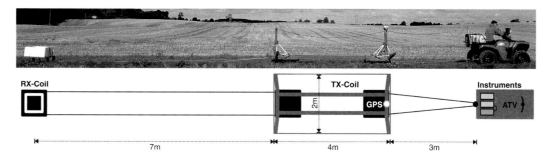

Figure 4.3 Ground-based tTEM system for mapping subsurface variability at the agricultural field scale. The all-terrain vehicle (ATV) tows the transmitter frame (Tx coil) and the receiver coil (Rx coil). (From Neven et al., 2021, distributed under a CC BY 4.0 license.)

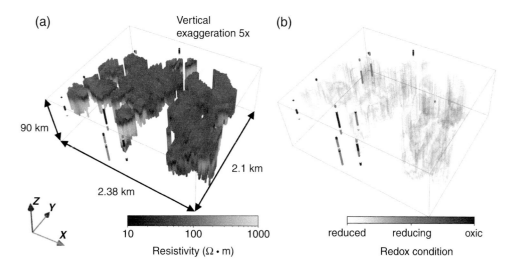

Figure 4.4 Three-dimensional visualization of towed transient electromagnetic resistivity (tTEM) data and redox boreholes. (a) Resistivity from tTEM survey; (b) redox conditions: reduced, reducing, and oxic from redox boreholes.

How can such mapping of the subsurface at the 100-m scale be done? Denmark has been at the forefront of developing geophysical imaging methods for mapping groundwater systems. One method is in the sky, the other on the surface. The so-called SkyTEM is an airborne geophysical survey for the acquisition of transient electromagnetic (TEM) data. It is like radar in the Antarctica case, just another physical wave form. However, airborne data do not provide enough detail (resolution) about the soil variation in a single agricultural plot. Therefore, a ground-based system (towed TEM or tTEM) has been designed that pulls the geophysical imaging device using a simple tractor; see Figure 4.3.

What information does this geophysical imaging provide us with? This is shown in Figure 4.4a. The specific case is at the west of Javngyde catchment in the central part of the

Jutland peninsula in Denmark, provided by the HydroGeophysics Group at Aarhus University and the Geological Survey of Denmark and Greenland. After data acquisition and processing, the result is, at each sounding location, a one-dimensional profile of the apparent resistivity of the soil, going down up to 100 m. So, it does not directly measure interfaces between oxic and reduced conditions. However, the resistivity variations are a function of the variation in soil properties, such as porosity, mineralogy, and lithology, which are all factors that may inform us about the redox conditions. Luckily, in the same area, we also have a few boreholes, see Figure 4.4b. In these boreholes, we do measure the required geochemical conditions. The problem, as often in the real world, is that these boreholes may not be overlapping with the tTEM survey. Hence, the following questions will be addressed in this chapter:

- Can we use the one-dimensional soundings to interpolate/simulate a three-dimensional model of resistivity in the subsurface?
- How can we correlate resistivity to redox potential? Is there any correlation?
- Can we use this potential to map the redox potential in three dimensions, thereby identifying agricultural areas with a good denitrification potential?

Taking all these questions together, this study supports a more sustainable agricultural regulation decision.

4.3   Concept Review

In this chapter we cover aspects of three-dimensional modeling. If you need a refresher, please consult the following sections:

- Section 5.3.2, where we discuss the relationship between various measures of correlations, such as the covariance and semi-variogram
- Section 5.3.3, where we cover properties of rotation in three-dimensional spaces
- Section 5.2.3, on Monte Carlo simulations

4.4   Spatial Analysis with Variograms

Variograms are the foundation of traditional geostatistics. They will help us in describing complex spatial variations using very few parameters. Variograms can be applied to any dataset, whether on a regular grid or unstructured data (data that do not follow a regular grid structure). Consider, for example, Figure 4.5, which shows various 100×100 km snapshots of the Arctic. A geomorphologist or glaciologist will describe for you in words what they observed. What we will do in this section is describe these images with a few quantities.

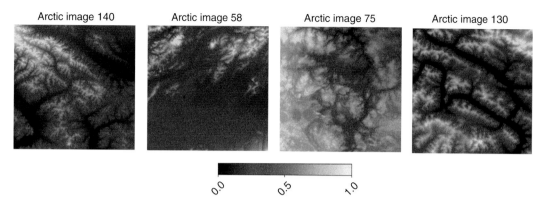

Figure 4.5 Four remote sensing images of the Arctic, scaled between zero and one to show spatial geomorphological patterns only. The image numbers are identification numbers (IDs) representing different locations. Each image covers an area of 100 by 100 km, on 200 × 200 grids with a resolution of 500 meters.

4.4.1 Calculating the Semi-Variogram in Two-Dimensional Space

Central to any spatial analysis is the notion of distance in space. The advantage of using a distance is that one is not married to a grid of values. Using distances, we can perform spatial analysis on regularly and irregularly spaced data. However, to start with, for ease of explanation, we will take regularly gridded data, such as in Figure 4.5. When we look at these images, we observe that color variations occur in well-defined regions; the images we are looking at are not pure randomness, we use the term "spatial correlation," which indicates all forms of spatial variability, but it has also very specific meaning. The question then is: How can we make "spatial correlation" quantitative?

Basic Concept

To do so, we start by studying the difference between two values in a grid, which are separated by a vector **h**. In Figure 4.6, we start with a vector that lies along the 45-degree direction and the distance is three pixels in the x-direction and three pixels in the y-direction, hence the length of this vector is $3\sqrt{2}$. Recall that a vector, starting at $(0, 0)$ in two dimensions, is uniquely defined by two values. Either you define an x and y coordinate or you define an angle together with a length. The vector here has a beginning and an end, which we call the head and tail values. If we look at the white vector, we observe that the head value = 1.5 and the tail value = 1.8. We have collected a pair of values (1.5, 1.8) along a vector **h**. However, one such pair taken at one specific location in the domain is not informative about spatial correlation in a specified domain. We'd like instead to know the typical difference between any two values along a vector **h** in the domain. To do so, we move the vector around the image collecting all possible pairs of observations, starting at the bottom left and ending at the top right. To get to a single measure, we calculate the average square difference between the two. We can put these words into an equation as follows:

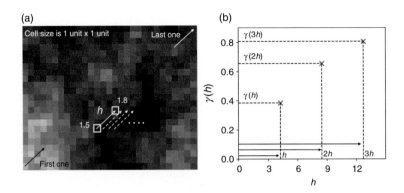

Figure 4.6 (a) Introducing the concept of variogram on a small regularly gridded dataset. (b) A variogram value is the square of the average difference between any two values at a given distance and direction h.

$$\gamma(\mathbf{h}) = \frac{1}{2N(\mathbf{h})} \sum_{n=1}^{N(\mathbf{h})} \Big(z(\mathbf{x}_i + \mathbf{h}) - z(\mathbf{x}_i) \Big)^2 \tag{4.1}$$

Let's explain this equation in more detail. We call the value at a grid location $\mathbf{x}_i$: $z(\mathbf{x}_i)$. Here we observe the introduction of a variable that is a function of space: z is a function of $\mathbf{x}_i$. Likewise, if we take the location $\mathbf{x}_i$ and add $\mathbf{h}$, we get $z(\mathbf{x}_i + \mathbf{h})$, the tail value. We make the square difference between the two: $(z(\mathbf{x}_i + \mathbf{h}) - z(\mathbf{x}_i))^2$. Now we do this for all the pairs we can possibly obtain: the number of pairs is $N(\mathbf{h})$. This number changes when the length of $\mathbf{h}$ changes. In fact, as $\mathbf{h}$ increases in length, you will get fewer pairs. Note also the $1/2$ at the front of the equation, which we will return to later.

In geostatistics, the function in Equation (4.1) is termed the semi-variogram. But because this is a lot of letters, geostatisticians will often just say "variogram." If we do the calculation for that $\mathbf{h}$ (length=$3\sqrt{2}$, 45-degree direction) we find:

$$\gamma(\mathbf{h}) = 0.39 \tag{4.2}$$

Before moving on: let's consider the following thought experiment: what if we now increase the length of $\mathbf{h}$ by a factor of two, three, four, etc., but keep the direction the same: What would happen to the $\gamma(\mathbf{h})$ function? If we look again at the image in Figure 4.6, we'd expect it to increase, because values further apart are likely to become increasingly different. Let's check this with actual calculation, in Figure 4.6. Indeed, a doubling of the length of $\mathbf{h}$ increases the variogram value to 0.65. A very important point to emphasize here is that the direction of the vector remains the same, here 45 degrees.

In Figure 4.6, we plot the length of a vector $\mathbf{h}$: ‖$\mathbf{h}$‖ versus the variogram value. Recall that we started with a vector of length $3\sqrt{2} = 4.24$, which we then multiply each time. We often replace the notation ‖$\mathbf{h}$‖ with the simpler notation length h. In geostatistics, we call this length the "lag distance." In these types of plots, which we will use many times, always remember that the x-axis is a distance and the y-axis is a measure of difference.

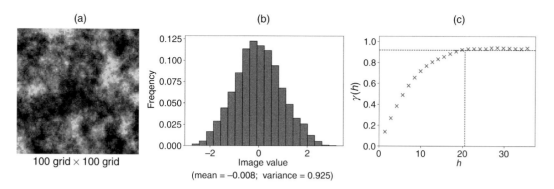

(a)

100 grid × 100 grid

(b)

Image value
(mean = −0.008; variance = 0.925)

(c)

Figure 4.7 (a) Two-dimensional image with 100×100 grid units, which is plotted as (b) a histogram, showing that the values have a standard Gaussian distribution; (c) the experimental variogram along the 45-degree direction, which increases with distance and then levels off.

We move now to a larger, still regularly gridded domain, repeating what we have done with **h** along the 45-degree direction; we calculate the variogram for many lag distances. Figure 4.7 shows that the experimental variogram along the 45-degree direction increases; however, at some point it levels off to a constant. This level here appears to be close to 1.

Recall that our target is to summarize what is observed in datasets with a small number of characteristics, i.e., scalars. We introduce two of these scalars:

• *range*: the length/lag distance at which the experimental variogram values level off
• *sill*: the value of that level

Because a range is a length (a distance) it is also termed "correlation length." If we look at Figure 4.7, we can guestimate that the *range* = 20 and the *sill* = 0.925. What does this range mean?

To understand this, look forward to Figure 4.9a: here, we draw circles around the patches we observe, and we can roughly attribute the average size of a patch with the value 20. This also means that if we take the center of such a circle and travel a distance of more than 20 grid sizes away, we expect to see very different "colors." For example, if we start with a particular color, we are likely to end up with a different color. This observation now also provides meaning to the sill value. If you look at the histogram (Figure 4.7b) of the values in this image, you will observe that the variance equals 0.925. Hence, here the sill equals the variance. This means that if we walk more than 20 grid sizes away, we notice that we can encounter all colors of the colorbar, or we observe the range of variation = variance.

The next step in this process is to change the direction of **h**, i.e., repeat the same exercise as before but in other directions. Figure 4.8 shows this by calculating the variogram along the −45-degree direction, which is orthogonal to the 45-degree direction. We notice that there is no significant difference between these variograms. In fact, as an exercise, you can do this for many orthogonal directions, and you would make the same observation. This is not really a surprise when looking at Figure 4.7a. Imagine this image represents topography, with valleys and hills.

Figure 4.8 Variograms in two directions: the 45-degree and −45-degree directions.

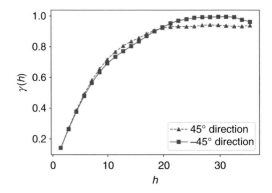

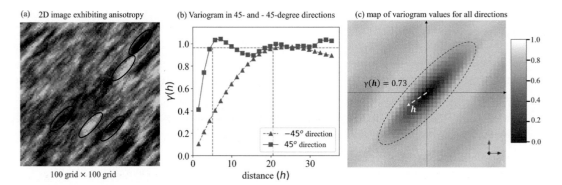

Figure 4.9 (a) Two-dimensional image with 100×100 grid units, showing anisotropy; (b) variograms along 45-degree and −45-degree directions; (c) map with variogram values for all directions centered at $\mathbf{h} = (0,0)$.

Imagine you stood on a hill; if you looked around in various directions, you would not see major differences in the style of topography, regardless of what location you were at. This notion is termed: isotropy (a word originated from ancient Greek, iso = equal; tropy = way).

Geometric Anisotropy

We move on to a new dataset to illustrate the next concept; see Figure 4.9. Visually, the main difference between Figure 4.7 and Figure 4.9 is that now features in the image are aligned along a certain direction. We can check this by calculating the variograms along different directions. Because of the the nature of the image, we take the 45-degree and −45-degree directions and we plot the two calculated variograms in Figure 4.9b. What is now striking is that the range is a function of the direction, along the 45 degree we have a range of 20, along the degree we have a range of 5, the sill remains close to 1 for both cases. This concept is termed: geometric anisotropy, the ranges change with direction.

For the dataset in Figure 4.9, it is visibly clear why this happens. For an intuitive understanding, imagine again Figure 4.9 is a topography; if you walk along the 45-degree direction, you will be walking mostly in valleys or on top of hills, it's an easy hike! If you walk in the −45-degree direction, your hike is tougher, you experience a lot of variation (or variance).

How do we turn these descriptive notions into a few simple scalars, which as you recall is our ultimate quest? To understand how this is done consider another thought experiment. What would happen to the range when we calculate the variogram for the direction between 45 and −45 degrees? We would expect that those ranges would be somewhere between 5 and 20 grid sizes. You will also notice that if you plot these ranges and their directions in a single graph, they pretty much align with an elliptical shape (Figure 4.9c). This brings us to the next step in our journey for simplification: geometric anisotropy is described by means of an ellipse. An ellipse in two dimensions is unique defined by

- *major range*: the length of the ellipse
- *minor range*: the width of the ellipse
- *azimuth/angle*: the orientation of the ellipse as measured from north to south (y-axis)

Figure 4.9 illustrates where exactly this ellipse comes from. In Figure 4.9a, we can distinguish patches of alike color that appear roughly elliptical in shape. Figure 4.9c is a little bit more tricky to get. Here, we plot all values of $\gamma(h)$ as a function of h in a two-dimensional map, centered at $\mathbf{h} - (0,0)$. Hence, if you take any direction from the origin and look up the variogram values in this map, you get a one-dimensional variogram. This plot also shows the symmetry of the variogram definition, namely that it does not matter if you go forward or backward along any particular direction.

Degree of Smoothness

Let's consider the three datasets/images shown in Figure 4.10a–c. No significant anisotropy is apparent, yet these images look very different. To get better insight into this particular spatial feature, let's focus on a small area. In Figure 4.10a, we observe that colors are only gradually changing. If we walked over this topography, we would find it pleasant and smooth. On going to Figures 4.10b and c, we see an increase in the roughness of the topography. What does this mean for the variogram? Smoothness is related to what happens over small distances, so we need to investigate the variogram for small lag distances h. Indeed, we notice a different change in the variogram for small distances. Geostatisticians have assigned particular mathematical functions to these different types of behavior at the origin (near 0). There are many types, but we will stick to the three most important types, which apply the vast majority of cases.

Their behavior near the origin is as follows:

- Gaussian type: the function goes as $1 - exp\left(-h^2\right)$ for small h
- spherical type: the function goes as h for small h
- exponential type: the function goes as $1 - exp(-h)$ for small h

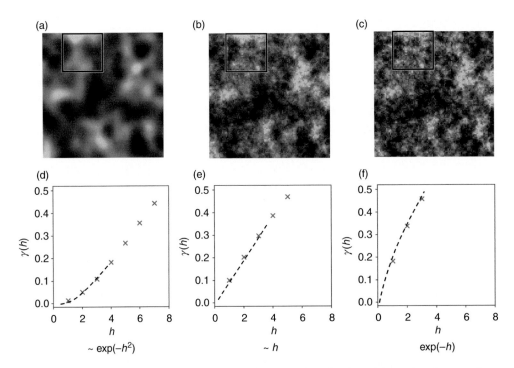

Figure 4.10 (a)–(c) Two-dimensional images with 100×100 grid units, with varying degrees of smoothness/roughness. (d)–(f) Their corresponding variograms investigated for small h.

Noise or Nugget Effect

We know from previous examples that $\gamma(0) = 0$. This is true: the average square difference between a value and itself is zero. However, that may not always appear to be the case. Consider the dataset in Figure 4.11. If we look at the variogram, there is a jump between the $h = 0$ and $h = 1$ pixel/cell. In fact, if we took the first three variogram values, we could extend them to the y-axis and observe an "apparent cut-off" at 0.25. In geostatistics, this apparent cut-off is termed the nugget effect; it is also called "noise" as in signal processing. The term noise, however, hides a more detailed story. Noise suggests some contamination of the signal. But the nugget effect tells a more interesting story. The nugget effect refers to a phenomenon observed in gold deposits, which often contain large gold nuggets. The largest gold nugget ever found the "Welcome Stranger," weights 97 kg! Consider a gold deposit like the one shown in Figure 4.12. When we drill for gold, we core the rock and hence we use a certain sample volume. Then, as common in all geosciences sampling, we also use a sample spacing. This means that, by chance, we may miss or hit a gold nugget. This shows up as a nugget (hit versus miss) in the variogram. What causes the nugget effect therefore is related to the size of what you are studying and the sample volume and distance between samples. Likewise, if samples are very large, the same problem occurs: we may not observe variation within the large sample, such as clustering of gold nuggets.

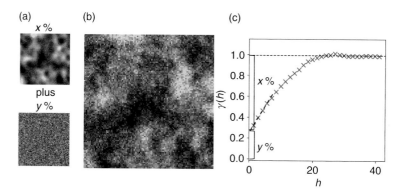

Figure 4.11 (a) Linear combination of two images, one of which is pure noise, resulting in image (b) with 100 ×100 grid units. (c) The variogram exhibits a nugget effect.

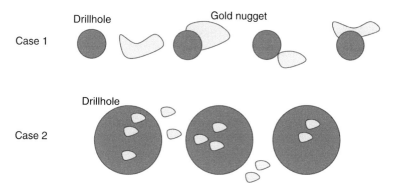

Figure 4.12 Two ways to create a nugget effect.

An interesting scalar to calculate is the ratio between the nugget effect ($y\%$) and the total variance ($x\%+y\%$); Figure 4.11c shows this to be 0.25 ($y\%$). What this means is that about 25% of the signal is "noise." Knowing the nugget effect helps us to remove noise from data, if that is desired.

Trends

Now that we have described variation at small distances, we move to larger distances. Figure 4.13 shows an image which has a well-defined "trend." A trend in geostatistics refers to the idea that some deterministic large-scale variation exists, meaning that in the image there is a clear change from low to high values when moving from the bottom left to the top right, despite the local variations. If we measure the differences in the values along this direction, they will keep on increasing, or, in terms of the variogram, it no longer has a sill. In the direction orthogonal to the trend, we still get the usual sill and range.

We will discuss in much greater detail in Section 4.5.5 how trends are treated in geostatistics, in particular when performing interpolation. What often happens is that a trend can be quite confidently estimated from data, even with relatively sparse datasets. If that is the case, the

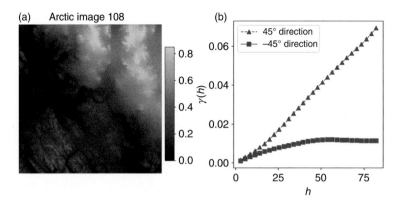

Figure 4.13 (a) Topography in an image (ID 108) from the Arctic dataset showing a systemic change in elevation. The image covers a 100×100 km area, on 200×200 grids. (b) The variogram in the direction of that change no longer exhibits a constant sill. The unit of distance (h) is 500 m.

deterministic trend can be subtracted from the data and we can focus our attention to smaller-scale residuals, resulting in a residual variogram.

A Variogram Interpretation Protocol for Two Dimensions

We now summarize what we have described so far. Keep in mind that the idea is to find a small number of scalars that can describe spatial variation in two dimensions. Here we describe an easy protocol (Box 4.1) that you can follow to summarize what you observe after calculating variograms.

Step 1: Calculate the histogram of your data and note the variance. The variance is going to be important in interpreting the sill of the variogram.

Step 2: Calculate the experimental variogram in a few directions, typically four to six directions. The variogram is symmetric with respect to 180-degree rotation, one option can be to take 0, 30, 60, 90, 120, 150 degrees. The angles in most software are measured from north to south, so zero means north to south.

Step 3: Determine the nugget effect. Knowing there is any noise or whether the total variation is not fully sampled is very important and has considerable consequences later when you will be using the variogram to perform interpolations.

Step 4: Determine the type of behavior at the origin: the smoothness. This will also have large effects in interpolations, so it is important to pay close attention to this.

Step 5: On the variogram plots, create a horizontal line that is equal to the variance; this may be indicative of a sill.

Step 6: If most of the variograms calculated over all directions reach, approximately, the same plateau (sill), you are ready to investigate whether you have isotropy or anisotropy. If you have anisotropy, then you need to mark off three values: the major range, the direction along which you can find this, and the minor range. The latter is now assumed to be orthogonal to the major range.

Box 4.1 Variogram interpretation protocol for two dimensions

- Calculate the data histogram and variance
- Calculate the experimental variogram in a few directions
- Determine the nugget effect
- Determine the type of smoothness at the origin
- Create a horizontal line that equals to the variance to find the sill
- Observe anisotropy
- Check if there is a trend in the data; subtract the trend and repeat the above steps if a significant trend presents

Plots to make

- Histogram of data
- Experimental variogram in a few directions (e.g. in directions of angles 0, 30, 60, 90, 120, 150 degrees)

Step 7: You may have a trend, which usually means that the variogram keeps increasing, or there is no clear sill. Identify this issue, possibly model the trend (see Section 4.5.5) and subtract it, then do steps 1 6.

There are also some things that you shouldn't do! We provide a number of important hints/tricks for avoiding beginners' mistakes in variogram modeling. The first one concerns interpreting the variogram in the context of interpolation (kriging). For example, you'd like to make a nice interpolated map based on some sparse sample data. As we will see in Section 4.5, we use the variogram in that context. In the context of interpolation, what matters most about the variogram is what happens at relatively short distances. It is very tempting to start focusing on the often wild fluctuations of the variogram for larger distances. And who wouldn't? It looks very interesting! Yet, as we will see, these fluctuations matter much less, for interpolation, as compared to shorter distances. What matters most is (1) the nugget effect, (2) the type of behavior at the origin, and (3) whether or not there is any significant anisotropy.

Also of importance is not to calculate the variogram for very large distances, more specifically, larger than half the size of the domain. We noticed before that as you increase the distances, you will get less pairs to calculate the variogram, so the values for large distances become less reliable. For very large distances you only compare values at the edges of the domain and leave out values in the middle. So, the variogram values you calculate are not really representative of the entire domain.

▶ Play **Video 08: Variogram** to study variograms of two- and three-dimensional images, and learn how to interpret variograms. This includes the tutorial on using SGEMS for empirical variogram calculation and variogram modeling.

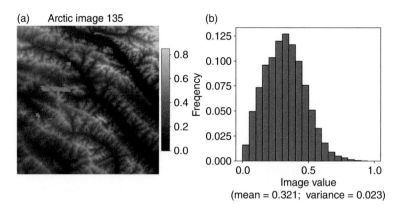

Figure 4.14 (a) Image (ID 135) of a 100 × 100 km area of the Arctic, on 200 x 200 grids. (b) A histogram of the data from this image. The topography has been scaled to values between 0 and 1.

Worked-Out Example: Calculating Variograms with Regularly Gridded Data of an Arctic Digital Elevation Model

Figure 4.14 shows an exposed area of the Arctic over a 100 × 100 km area. As mentioned before, the nature of the Arctic bedrock may be helpful in understanding the Antarctic. Variograms are ways to describe spatial variability, and hence the ability to display variograms of both the Arctic and Antarctic is useful in gaining understanding about similarities and differences. Just visually, one can appreciate the stunning spatial variation present in the image, with both large and smaller-scale features clearly prominent. But before getting too excited, we run the variogram protocol:

Step 1: Calculate the histogram of your data and note the variance. This variance here equals 0.023. Note that we scaled the digital elevation model values to be between 0 and 1 for easy interpretation. This variance will be important in Step 6.

Step 2: Calculate the experimental variogram in a few directions. Here we calculated the variogram in the 0-, 30-, 45-, 60-, 90-, 120-, 135-, 150-degree directions from north to south.

Step 3: Determine the nugget effect. We zoom into any variogram and notice there is no nugget effect. This is expected, it is a glaciated topography, so no rough edges were missed in the data.

Step 4: Determine the type of behavior at the origin: the smoothness, in other words. This is Gaussian type, as shown in the inset of Figure 4.15.

Step 5: On the variogram plots, create a horizontal line that is equal to the variance; this may be indicative of a sill. We notice now that many of the directions plotted reach the sill, except for the 135-degree direction.

Step 6: Check for anisotropy. Anisotropy is clearly present; in fact, one direction, the 135-degree anisotropy is so strong that the variogram along that direction does not fully reach the sill. This is in fact quite a common observation on many phenomena, certainly those that have strong continuous features (e.g., glaciated valleys, layers). This is also means that, in this case, the 135-degree direction is the glaciated valley direction, clearly visible in the dataset.

Step 7: There is no significant trend present in this dataset.

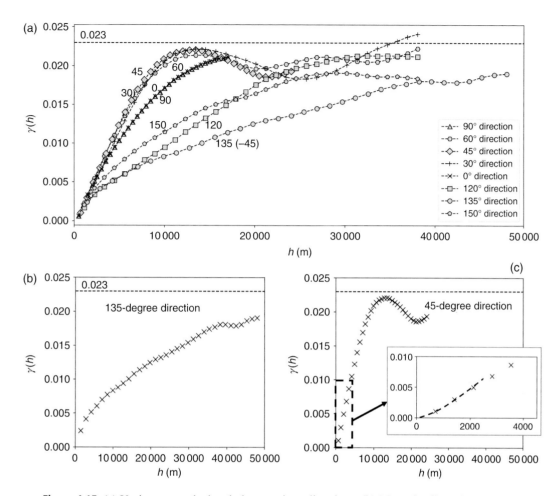

Figure 4.15 (a) Variograms calculated along various directions. (b) More detail on the 135- and 45-degree directions.

In the next section, we will be calculating variograms for the bedrock of the Antarctic. Some of the key takeaways here are that:

- the bedrock is smooth (Gaussian type) due to glaciation
- the glaciated channels and their direction show up as geometric anisotropy, sometimes with very long ranges in the major direction
- the range in the minor direction indicates the width of the valleys, here around 10 000 m
- there is no nugget effect

Calculating Experimental Variograms with Irregularly Spaced Data

Many datasets do not come with a grid; samples are recorded at irregularly spaced locations. A good thing about the variogram is that we can also calculate it, with some

Figure 4.16 Collection of irregularly space samples. A black and white version of this figure will appear in some formats. For the color version, refer to the plate section.

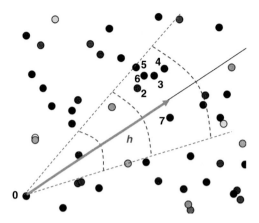

additional effort, on any data configuration. This is a very useful property of the variogram because now we can calculate and compare variograms of datasets that have any amount of irregularity in sampling.

The main problem with irregularly spaced data is that we can no longer apply the idea illustrated in Figure 4.16: collecting pairs of observation that lie exactly along a given vector **h**. We need to introduce some additional tolerances; we can no longer be extremely rigorous about **h**.

There are two ways to create a tolerance on a vector, one is an angle tolerance, the other is a length tolerance. Figure 4.16 shows how this works. Consider the datapoint denoted as "0." To calculate the variogram we need to seek other values in the dataset that lie along a given direction, here 45 degrees and with varying length, in order to calculate the variogram along the 45-degree direction. What we do here is to take a tolerance on the angle of direction and a tolerance on the distance. This creates a pizza-slice shape with compartments along its axis. We now assume that each point within these compartments is at the same distance and direction from point 0. When we move this pizza slice around, centering it on each data point, we will get many pairs of samples that can be used to calculate $\gamma(\mathbf{h})$, where **h** is now the center of each compartment.

You can see that you need to make a compromise now: if you make the angle too wide, then you will mix too many data from different directions, and so your experimental variogram is no longer reflective of variation along the 45-degree direction. If you make it too narrow, you may have too few points and the variogram fluctuates a lot, which makes it hard to interpret. In other words, you will need to try these tolerances by trial and error.

Worked-Out Example: Variograms of Radar Flightline Data

Here, we work out, in great detail, how to calculate variograms on irregular data and how to use the above protocol for interpreting the variogram. The example concerns an area of the radar flightline data on the Thwaites glacier, shown in Figure 4.17.

The data are clearly irregular, with varying flightline density over the domain. Also, just by eye, we can distinguish areas of high altitude (north and south in the domain) as well as low

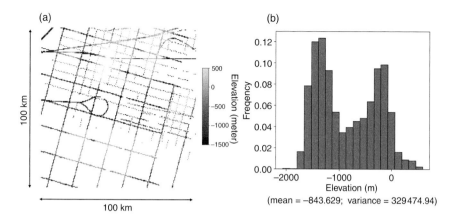

Figure 4.17 (a) Flightlines over a portion of the Thwaites glacier showing irregularly spaced data. (b) The histogram is bimodal reflecting the nature of glaciated valleys alternated with mountains.

altitude, possible a glaciated valley. We start by calculating the variogram in the east–west direction. To do so, we need to specify tolerances, namely, angle and distance tolerance. For angle tolerance, it is recommended to start with about 20 degrees, although you may need to take a higher tolerance when fewer data are available than in this case. For lag-distance tolerance, we need to understand better how far apart data points are located from each other. In this dataset, we have flightlines that are about 15 km apart, and some others 5 km apart. In addition, the data along the flightlines are as close as 50 m. This poses a challenge, because tolerances that are too large would ignore the short-scale variation along flightline, while if they are too small excessive computation may be required and you may possibly obtain very noise variograms.

Good practice is to start out with one or two directions. It is logical to start with directions we know are interesting, here north–south and east–west. Don't aim for perfection when calculating your first variograms. The first decision is around the lag distance (the length of h) and how many lags to calculate. The second is on lag and angle tolerance. We start with a lag distance of 300 m. With some trial and error, you'll find that a lag tolerance of 150 m (half the lag distance, which is good choice). An angle tolerance of 20 degrees gives the experimental variograms along eight different directions; see Figure 4.18.

Here is the variogram interpretation protocol with the above example:

- The variance is 329 546.
- Variograms are calculating along eight directions, including the important directions of north–south, east–west.
- There is no nugget effect.
- The behavior at the origin is Gaussian.
- None of the variograms plateau around 329 546.

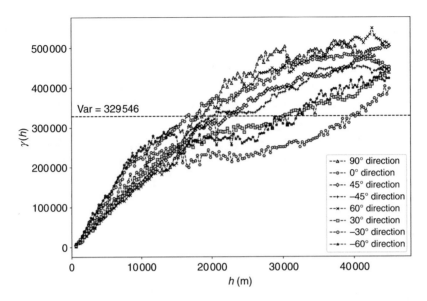

Figure 4.18 Variograms along various directions.

- We observe all variograms exceeding the variance, but some variograms, such as the north–south variogram, declines again around 40 000 m.
- The dominant trend of mountain–valley–mountain that runs north–south has a significant effect on the variogram. This type of trend in geostatistics is also termed the "hole effect." The "hole" here is a valley in between two mountains. If we had only a mountain–valley type trend, all the variograms would keep increasing (increasing difference). The addition of a mountain far across a valley creates more similarity over large distance (mountain to mountain), hence the decrease.

4.4.2 Limitations of Variograms

There are many more ways to analyze spatial variability in two-dimensional images than using variograms. Computer vision is an area of computer science that extracts information from images. For example, it can look at images of the Moon and extract craters from it, then measure the size of craters and plot the results in a histogram. In that setting, a variogram would not be useful. The anisotropy, range, nugget parameters therefore measure geometries to some degree, but not fully or exhaustively. An example of this is shown in Figure 4.19. These landscapes are clearly very different, yet they have the same variogram. Variograms therefore are not very well suited to describe very specific structures or geometries. We will revisit this issue in Section 4.7.

Where variograms excel is in the quantification of spatial variability when you have sparse and/or irregularly spaced data. In those cases, many computer vision methods become difficult to apply.

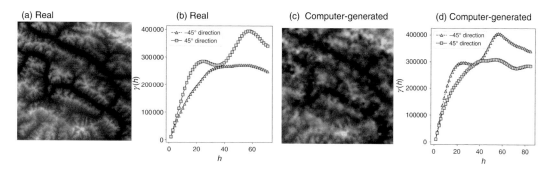

Figure 4.19 (a) Image (100×100 km) and (b) variogram of a real Arctic topography versus (c) a computer-generated image and (d) corresponding variogram. The unit of distance (h) is 500 m.

4.4.3 What Have We Learned So Far?

- We have learned about variograms, which describe spatial variations by calculating average square differences between values in a spatial dataset as function of distance.
- We have learned that a very basic quantitative description of a two-dimensional spatial dataset may involve as few as six parameters: the sill, the range in two orthogonal directions, the azimuth angle, the nugget effect, and the behavior for small distances.
- We have learned that variograms are a powerful tool to calculate spatial variation for irregularly spaced data.

4.5   Interpolation with Kriging

Once we have measured the spatial variations using variograms, we can fill (interpolate) the spaces of the irregularly measured data to get a map in high resolution, for example, filling the gaps between the flightlines (see Figure 4.17). In this section, we start performing interpolation with a linear regression technique, kriging.

4.5.1 What Is Kriging?

Kriging, a method of spatial estimation and interpolation, is how geostatistics took off. As mentioned above, the name kriging refers to a South African mining engineer, Danie Krige (Cressie, 1990), while the technique was fully developed by a French mathematician, George Matheron (1969, 1971). As shown in Figure 4.20, the goal of kriging is to take a point set and turn it into an interpolated map. If we look carefully at this map, we observe a few things:

- The map (Figure 4.20b) does not contain artifacts, meaning you don't see the (line) geometry of the data.
- The map (Figure 4.20b) is smooth.
- The interpolation is exact, meaning that at the data locations the actual data are matched perfectly.

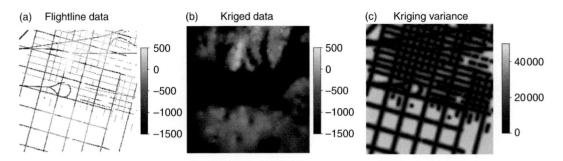

(a) Flightline data (b) Kriged data (c) Kriging variance

Figure 4.20 (a) Flightline data, (b) kriging interpolation, and (c) kriging variance of the interpolation.

- A second map is shown (Figure 4.20c), termed kriging variance, which reflects the confidence in the interpolation. At locations of sample data, we have 100% confidence, and that confidence drops as we move away from the sample data.

Technically speaking, kriging is the best linear unbiased estimator (BLUE). But in this book, we will skip any theoretical development or derivations, and instead try to understand what kriging does, and how the variogram is used in that context. There are many flavors of kriging, but they all rely on very similar principles. Kriging provides an estimate of the value at an unsampled location, based on nearby sample values. When we perform this estimation at a grid of location, we get the smooth map shown in Figure 4.20b.

4.5.2 What Are the Important Principles Used in Kriging?

In kriging, see Figure 4.21, we estimate the value at an unsampled location by giving weights to nearby samples, thereby constructing the following estimator:

$$\hat{z}(\mathbf{x}) = \sum_{\alpha=1}^{n} \lambda_{\alpha}(\mathbf{x})z(\mathbf{x}_{\alpha}) \tag{4.3}$$

where $\hat{z}(\mathbf{x})$ is an estimator at the unsampled location $\mathbf{x}$. It is not equal to the true value, so we put a hat on it to differentiate it from sample data, which are considered the true values at their location. The value $z(\mathbf{x}_{\alpha})$ is the nearby sample value at location $\mathbf{x}_{\alpha}$; λ_{α} is the weight. The question now is how to determine the weights. Instead of deriving equations, we state two principles that are used in the derivation:

- *Principle one*: Data that are closer to the location to be estimated should get a larger weight, if the closeness results in a large correlation between the sample and the unknown.
- *Principle two*: Data that are close to each other should share weights (thus smaller weights) if they are redundant (have very similar values) with each other.

Both principles require a definition of "closeness." This is what we will focus on. To measure closeness in space, we typically use the Euclidean distance. Consider the situation in

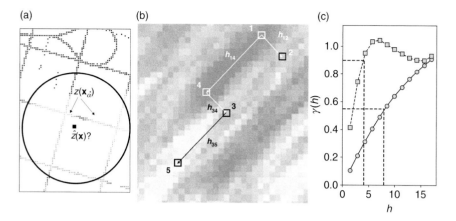

Figure 4.21 (a) Kriging estimates the value at an unsampled location from nearby data. (b) The Euclidean distance between these five samples is not informative of how correlated the samples are. (c) The variogram of (b) along the major and minor directions; the unit of distance (h) is 500 m.

Figure 4.21b, where we have an unknown truth (the background map) and five samples taken from it.

The Euclidean distance is the length of the vector between any two samples. The length of a vector $\mathbf{h}_{ij}$ has the notation $\|\mathbf{h}_{ij}\|$. Based on the data configuration, we can make the following statements in terms of length of the $\mathbf{h}$ vectors:

$$\begin{aligned} \|\mathbf{h}_{12}\| &< \|\mathbf{h}_{14}\| \\ \|\mathbf{h}_{34}\| &< \|\mathbf{h}_{35}\| \end{aligned} \tag{4.4}$$

However, we also notice that samples 1 and 4 lie along an area with strong spatial correlation, while samples 1 and 2 lie across it. If we have this knowledge, we expect samples 1 and 4 to be similar in value, while samples 1 and 2 would be very different. While we do not know the underlying truth, through proper variogram analysis we have some idea of the spatial variation of that truth. Hence, we may know the following:

$$\|\mathbf{h}_{12}\| < \|\mathbf{h}_{14}\| \quad but \quad \gamma(\|\mathbf{h}_{12}\|) > \gamma(\|\mathbf{h}_{14}\|)$$

Figure 4.21 illustrates why this is the case. Similarly:

$$\|\mathbf{h}_{34}\| < \|\mathbf{h}_{35}\| \quad but \quad \gamma(\|\mathbf{h}_{34}\|) > \gamma(\|\mathbf{h}_{35}\|)$$

In other words, the variogram is a measure of closeness that included both the Euclidean distance and the spatial correlation.

Let's turn a practical setting in Figure 4.22, where we need to estimate the value at location 0 from five sample values. In principle one, we consider closeness between samples and the unknown. The unknown is indexed with 0. If we have five samples and one unknown, we can gather all variogram information into a single vector:

Figure 4.22 We can define two types of distances: the distance between locations of the data and the location of the value to be estimated, and distance between data locations.

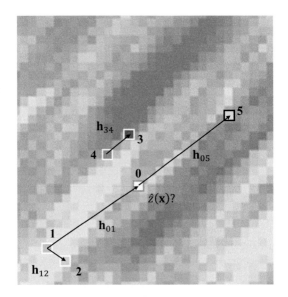

$$\begin{pmatrix} \gamma(\mathbf{h}_{01}) \\ \gamma(\mathbf{h}_{02}) \\ \gamma(\mathbf{h}_{03}) \\ \gamma(\mathbf{h}_{04}) \\ \gamma(\mathbf{h}_{05}) \end{pmatrix} \tag{4.5}$$

In principle two we need to compare sample values; this means creating a table or matrix that contains variogram values as follows:

$$\begin{pmatrix} 0 & \gamma(\mathbf{h}_{12}) & \gamma(\mathbf{h}_{13}) & \gamma(\mathbf{h}_{14}) & \gamma(\mathbf{h}_{15}) \\ \gamma(\mathbf{h}_{12}) & 0 & \gamma(\mathbf{h}_{23}) & \gamma(\mathbf{h}_{24}) & \gamma(\mathbf{h}_{25}) \\ \gamma(\mathbf{h}_{13}) & \gamma(\mathbf{h}_{23}) & 0 & \gamma(\mathbf{h}_{34}) & \gamma(\mathbf{h}_{35}) \\ \gamma(\mathbf{h}_{14}) & \gamma(\mathbf{h}_{24}) & \gamma(\mathbf{h}_{34}) & 0 & \gamma(\mathbf{h}_{35}) \\ \gamma(\mathbf{h}_{15}) & \gamma(\mathbf{h}_{25}) & \gamma(\mathbf{h}_{35}) & \gamma(\mathbf{h}_{35}) & 0 \end{pmatrix} \tag{4.6}$$

4.5.3 Spatial Covariance

Before getting into the specific kriging methods in the next section, we need to first introduce the concept of the spatial covariance function $C(h)$. *Similar to the format in Eq.4.7.* The definition of this function is as follows:

$$C(\mathbf{h}) = C_0 - \gamma(\mathbf{h}); \ C_0 = C(\mathbf{0}) = \text{variance} \tag{4.7}$$

This means that when the variogram increases, the spatial covariance decreases; see Figure 4.23. Why do we need this new concept? To understand this, let's revisit how the variogram was calculated. Variogram calculations require finding pairs of data in the domain that are a vector $\mathbf{h}$ apart, then we calculate their square distance:

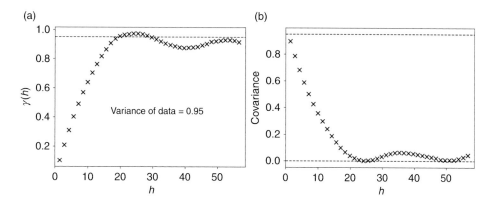

Figure 4.23 Comparing (a) the variogram of a dataset with (b) the spatial covariance of that same dataset. The unit of distance (h) is 500 m.

$$\Big(z(\mathbf{x}_i + \mathbf{h}) - z(\mathbf{x}_i)\Big)^2 \tag{4.8}$$

Averaging this and dividing by two provided the variogram value $\gamma(\mathbf{h})$ for that $\mathbf{h}$. To calculate the spatial covariance, we do the following (skipping the derivation):

$$\Big(z(\mathbf{x}_i + \mathbf{h}) - m\Big)\Big(z(\mathbf{x}_i) - m\Big) \tag{4.9}$$

For each value, we first subtract the mean m, then we take the product. Like the variogram we calculate the average sum (but we do not square and or divide by two):

$$C(\mathbf{h}) = \frac{1}{N(\mathbf{h})} \sum_{n=1}^{N(\mathbf{h})} \Big(z(\mathbf{x}_i + \mathbf{h}) - m\Big)\Big(z(\mathbf{x}_i) - m\Big) \tag{4.10}$$

Let's analyze this a bit further. When we increase $\mathbf{h}$, we expect larger differences, which in terms of covariance means a smaller sum. You can understand that by thinking about small $\mathbf{h}$, for those cases $z(\mathbf{x}_i + \mathbf{h}) - m$ and $z(\mathbf{x}_i) - m$ are similar, so their product on average will be larger.

The reason why geostatisticians prefer variograms is because their calculations do not involve the mean m of the data. That mean value m may not be very reliable in cases with a trend: the variogram avoids this issue, the covariance does not. However, the kriging method, which we discuss in the next section, requires knowing the spatial covariance, more specifically, in our previous five-sample problem, we need to know:

$$K = \begin{pmatrix} C_0 & C_0 - \gamma(\mathbf{h}_{12}) & C_0 - \gamma(\mathbf{h}_{13}) & C_0 - \gamma(\mathbf{h}_{14}) & C_0 - \gamma(\mathbf{h}_{15}) \\ C_0 - \gamma(\mathbf{h}_{12}) & C_0 & C_0 - \gamma(\mathbf{h}_{23}) & C_0 - \gamma(\mathbf{h}_{24}) & C_0 - \gamma(\mathbf{h}_{25}) \\ C_0 - \gamma(\mathbf{h}_{13}) & C_0 - \gamma(\mathbf{h}_{23}) & C_0 & C_0 - \gamma(\mathbf{h}_{34}) & C_0 - \gamma(\mathbf{h}_{35}) \\ C_0 - \gamma(\mathbf{h}_{14}) & C_0 - \gamma(\mathbf{h}_{24}) & C_0 - \gamma(\mathbf{h}_{34}) & C_0 & C_0 - \gamma(\mathbf{h}_{45}) \\ C_0 - \gamma(\mathbf{h}_{15}) & C_0 - \gamma(\mathbf{h}_{25}) & C_0 - \gamma(\mathbf{h}_{35}) & C_0 - \gamma(\mathbf{h}_{45}) & C_0 \end{pmatrix} \tag{4.11}$$

and

$$k = \begin{pmatrix} C_0 - \gamma(\mathbf{h}_{01}) \\ C_0 - \gamma(\mathbf{h}_{02}) \\ C_0 - \gamma(\mathbf{h}_{03}) \\ C_0 - \gamma(\mathbf{h}_{04}) \\ C_0 - \gamma(\mathbf{h}_{05}) \end{pmatrix} \qquad (4.12)$$

4.5.4   Simple versus Ordinary Kriging

So far, we have outlined some important principles that underlly any spatial interpolation. Now, we put these in to practice, starting with the simplest interpolation method, called "simple kriging."

Simple Kriging

We keep to our promise of not deriving equations, relying on other textbooks for this, and instead emphasizing the intuition behind these equations. In simple kriging we assume that the mean of the phenomenon being studied is constant and known to be a value m. In simple kriging, the weights can be determined as follows:

$$\text{Solve } K\lambda = \mathbf{k} \qquad (4.13)$$

Then, the estimator of the unknown based on five samples in Figure 4.22 equals

$$\hat{z}(\mathbf{x}) - m = \sum_{\alpha=1}^{5} \lambda_\alpha \left(z(\mathbf{x}_\alpha) - m \right) \qquad (4.14)$$

We can use this equation for any number of data, and not just five, which is what we will discuss next.

Elements Needed for Simple Kriging   Regardless of what software package you use for simple kriging, four main elements need to be specified:

- the sample data
- the mean
- the variogram
- the grid on which to interpolate

Typically, we estimate the mean by the arithmetic mean of the sample data. For the variogram, the software will ask for a "variogram model." In this book, we skip variogram modeling since this is a more advanced topic. A variogram model is a mathematical function that has several free parameters. We have covered the minimum amount of variogram parameters already in Section 4.4. In other words, if you have run the variogram modeling protocol, and estimated the required parameters, you have a basic variogram model, and the parameters can be directly used in the software. As a reminder, these parameters are (in two dimensions):

- variance
- nugget
- type (Gaussian, spherical, or exponential)
- major/minor range, azimuth angle

Making Simple Kriging Maps   Because kriging relies on distance only, we can use it to interpolate any location in two-dimensional space, either on the unstructured data or a regular grid. However, more typically, kriging is performed on a regular grid. We first create an empty grid with user-specified grid spacing, then we perform kriging at every location of the grid. We can be clever about doing this efficiently. The matrix K is not a function of the locations we are seeking to perform interpolation, only a function of the sample data. That mean that we can re-use K^{-1}. What changes each time is the right-hand-side vector.

Consider an example modified from the dataset in Figure 4.17, shown in Figure 4.24. In the modified dataset, we have removed the trend of the data. We will discuss how to do this in Section 4.5.5. What we are left with is a dataset that has a mean very close to 0 (see the histogram). Comparing the histogram in Figure 4.17b with Figure 4.24b, we also notice that the bimodal histogram now becomes a unimodal one. Figure 4.24c shows the variogram of this modified data, which looks very different from the variogram in Figure 4.17. Two main differences are observed:

- all variograms reach the same sill
- the strong anisotropy now becomes a case that is almost isotropic

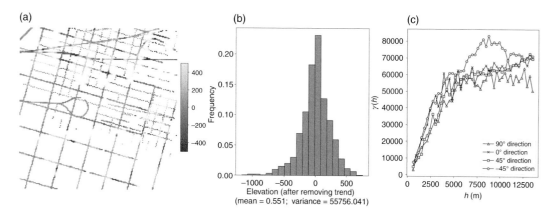

Figure 4.24 (a) Flightline dataset of Figure 4.17 that has the trend removed, resulting in (b) a unimodal histogram and (c) variograms that reach sill in all directions.

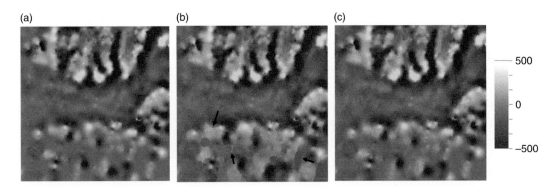

Figure 4.25 (a) Global simple kriging versus (b) artifacts (shown by arrows) that appear when performing local ordinary kriging with a fixed search neighborhood. (c) Ordinary kriging with a fixed number of data helps to mitigate the artifacts. Each box is 100 ×100 km, with a resolution of 500 m.

Applying the variogram protocol to the variograms in Figure 4.24c, we obtain

- variance = 550 000
- nugget = 0
- type = Exponential
- minor = major range = 6000; because the anisotropy is very minor, in this case, you may use an isotropic variogram, it will not greatly affect the kriging map

▶ Play **Video 9: Kriging** to interpolate spatial data using kriging. This includes the tutorial on using SGEMS for interpolation.

Stationarity of the Mean   Previously, we said that simple kriging assumes a "constant and known mean." This is a mathematical assumption used to derive the equations above but remains difficult to test quantitatively with actual data. This assumption is also termed "stationarity of the mean." What is assumed here is that there is no significant systematic variation in the average of the phenomenon. We assume the phenomenon can be adequately described by a constant mean with some correlated fluctuations around it. We also assume that by averaging the sample data we can estimate this constant mean confidently. The word "confident" is used more qualitatively than quantitatively in this setting.

A better and more practical way to think about this is to ask: What if that assumption is wrong? To understand this sensitivity, let's rewrite the estimator as follows:

$$\hat{z}(\mathbf{x}) = \left(1 - \sum_{\alpha=1}^{n}\lambda_\alpha\right)m + \sum_{\alpha=1}^{n}\lambda_\alpha z(\mathbf{x}_\alpha) \tag{4.15}$$

In this formulation, the mean looks like just another sample value, and the weight assigned to it is $\left(1 - \sum_{\alpha=1}^{n}\lambda_\alpha\right)$. So, in a situation with very few data, or when data have very little correlation with the unknown, the mean gets most of the weight. In areas with very few data, the estimator gets pulled towards the mean, and when we are extrapolating, we find the estimator is equal to the mean.

You can now see that, with only sparse data, the mean becomes very important. But then a problem arises: with sparse data we really don't know the mean very well, and we also don't know that it is constant over the domain. One solution to this problem is ordinary kriging with a local neighborhood search, or local ordinary kriging.

Ordinary Kriging

Simple kriging is rarely applicable in real settings, it assumes you have a lot of data and that fluctuations are around a constant known mean, which isn't so interesting. We will make two modifications: the first one is to get rid of the constant mean assumption, the second is to use only data near the location to be estimated.

A very easy way to get rid of the mean is to put a constraint on the weights, namely forcing the weights to sum to one.

$$\text{force} \sum_{\alpha=1}^{n} \lambda_\alpha = 1 \;\Rightarrow\; \hat{z}(\mathbf{x}) = \sum_{\alpha=1}^{n} \lambda_\alpha z(\mathbf{x}_\alpha) \tag{4.16}$$

which means that we need to solve a linear system of equations with a linear equality constraint:

$$\text{Solve } K\lambda = \mathbf{k} \text{ subject to } \lambda^T 1 = 1 \tag{4.17}$$

This has an analytical solution in linear algebra. Solving the weights this way is termed ordinary kriging.

Local Ordinary Kriging

So far, we have used all the data all the time, for all locations where we perform interpolation/estimations. This idea is called "global kriging." If we acknowledge that the mean is not really constant, then perhaps using all the data all the time is not such a good idea. Instead, we propose local kriging, which is a form of kriging where only data near to the location to be estimated are used. Indeed, why would you use flightline data 500 miles away from the location you want to interpolate? There are two reasons to promote this idea:

- The kriging matrix is of smaller size. For example, in the Antarctica case study, we have 4601 data (samples) of elevation; hence, we need to calculate K^{-1} for a very big matrix, which is not a good idea. Not only is this a computational problem, but matrices of that size may be close to singular.
- Data far away are not informative (they are not correlated) to what we are estimating locally.

Therefore, most kriging programs will use what is termed a local search neighborhood. When arriving at any location in the grid, the kriging code searches for sample values close by. This

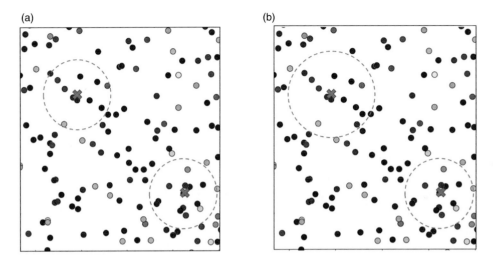

Figure 4.26 Two types of neighborhood definition: (a) with constant radius search neighborhood; (b) with a constant number of samples (14) in the neighborhood. A black and white version of this figure will appear in some formats. For the color version, refer to the plate section.

does require a definition of closeness that needs to be use-specified. Two ways to do this are as follows:

- Specify a search ellipsoid, as shown in Figure 4.26a. The definition of the search ellipsoid can be informed by the anisotropy of the data.
- Specify a minimum number of sample values, as shown in Figure 4.26b.

It is important to do this with care, as the use of these local neighborhoods may incur artifacts. This is shown in Figure 4.25b when we use local ordinary kriging. In the case of defining neighborhoods using local search ellipses (here a circle of radius 5000 m), we find clear artifacts that reflect the geometry of the search neighborhood. This is not the case when using a fixed number of samples (Figure 4.25c).

Kriging Variance

In previous chapters we covered how estimates are often accompanied by some measure of confidence. The estimate of the mean of a population from limited sample data is a function of two elements:

- How many samples do we have?
- What is the variance of the population?

We discussed a bootstrap method to calculate measure of confidence, such as quantiles and confidence intervals. We can't apply traditional bootstraps to spatial cases. In such cases, we

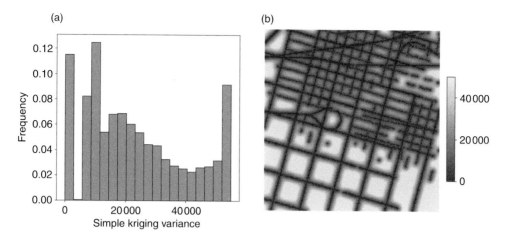

Figure 4.27 (a) Histogram and (b) map of the simple kriging variance.

have only one sample per location, so there is not much to bootstrap with. Luckily, in kriging, there is an analytical expression for the estimation variance, which tells us not exactly what the error is, but its variance; in other words, it provides us with some magnitude of the error, for example 1, 10, or 100.

The exact expression for the simple kriging variance is:

$$var(\mathbf{x}) = var(z) - \sum_{\alpha=1}^{n} \lambda_\alpha \Big(var(z) - \gamma(\mathbf{h}_{0\alpha}) \Big) \tag{4.18}$$

This expression has three components, the variance of the data (*var*), the kriging weights, and the variogram value between the data and the unknown. If there are no data, or all the weights are zero, then the kriging variance is just the variance of the data. When the weights become progressively larger (the data become more informative about what we are estimating), we subtract from the variance, and hence we become more certain about the true value. Looking at Figure 4.27, we observe how areas with dense data have a lower kriging variance.

4.5.5   Kriging in the Presence of a Trend

In the above section, we used simple and ordinary kriging to interpolate data. The trend in the data was removed. These kriging methods work well after removal of the trend, but less well when applied to the actual trended data. If you do this, you may find that the interpolation violates the "stationarity of mean" assumption. This is an important problem that we need to study.

A Geostatistical Model That Includes a Deterministic Trend

In many geoscientific datasets, a very clear trend is present in the data. For example, these trends can be due to seasonal changes. In the flightline data, we notice, just by plotting the data (Figure 4.25), a trend of mountain–valley–mountain. When such trend is clear, geostatistics offers techniques to incorporate this directly, improving the predictions and/ or interpolations. Underlying these techniques is a model of how geostatistics views variation in spatio-temporal data. We assume that a deterministic but unknown trend exists and that, on top of this trend, correlated fluctuations are added. In terms of an equation this entails:

$$Z(\mathbf{x}) = m(\mathbf{x}) + R(\mathbf{x}) \tag{4.19}$$

This is a model, it is not reality, but it can work for many real datasets to improve predictions. Notice the notation with small and capital letters. The $m(\mathbf{x})$ is deterministic in the sense that we can assume it can be estimated quite confidently from data, while the residual part, $R(\mathbf{x})$, is much more uncertain; the residual part captures variations at shorter scale. With this model, geostatisticians have derived several other flavors of kriging. In this book, we will only cover kriging with a locally varying mean. One component of that method is the trend function analysis.

Trend Function Analysis

The aim of trend function (sometimes called trend surface) analysis is to smooth the short-scale variation out, and retain only the large-scale variation or trend, or, according to the geostatistical model, $m(\mathbf{x})$. The end result is a deterministic, single trend function. The goal of trend function analysis is very different from the goal of kriging. In kriging, we interpolate the data and we get a measure of uncertainty on Z, the total signal. In trend function analysis, no such measure of uncertainty is generated (we aim for a trend we are confident in), plus we are interested only in $m(\mathbf{x})$, not $R(\mathbf{x})$. A very popular method for trend function analysis is radial basis function (RBF) smoothing. The idea looks very similar to kriging, so let's reframe kriging in words:

$$\hat{z}(\mathbf{x}) = \sum_{\alpha=1}^{n} \lambda_\alpha z(\mathbf{x}_\alpha) = \text{linear combination of sample data}$$

In RBF smoothing, we take the following linear combination

$$\hat{m}(\mathbf{x}) = \sum_{\alpha=1}^{n} \lambda_\alpha f(\|\mathbf{x}_\alpha - \mathbf{x}\|)$$

$$= \text{linear combination of functions } f \text{ centered on data locations}$$

where λ_α is again the weight for the linear combination, $f(\|\mathbf{x}_\alpha - \mathbf{x}\|)$ is a function f centered around the data location $\mathbf{x}_\alpha$, $\|\mathbf{x}_\alpha - \mathbf{x}\|$ is the distance between any location $\mathbf{x}$ and the location of data $\mathbf{x}_\alpha$. In this way, the function f is extended in a radial direction away from $\mathbf{x}_\alpha$, hence the name "radial basis."

In Figure 4.28, we take the following cubic function:

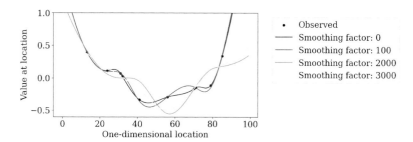

Figure 4.28 RBF smoothing using cubic functions with several smoothing factors. The aim is to smooth the data points (dots). For smoothing factor zero, we get perfect interpolation.

$$f(\|\mathbf{x}_\alpha - \mathbf{x}\|) = (\|\mathbf{x}_\alpha - \mathbf{x}\|)^3 \tag{4.20}$$

If we apply this method as it is (Smoothing factor = 0)., we notice that the RBF method interpolates the data perfectly. That is not what we want for kriging with trend; instead, we want to find the trend in this simple dataset. To do so we need to allow for some error in the interpolation. This amount of error is termed the smoothing factor. The higher this factor, the more error we allow, and hence the smoother the curve becomes, as shown in Figure 4.28.

At this point you may wonder why we cannot use the RBF smoothing with a smoothing factor of zero instead of kriging, as it seems so much easier! Indeed, RBF interpolation is a valid way to interpolate but it has a number of disadvantages:

- In RBF interpolation, you need to choose the radial basis function. In kriging you estimate variograms from the dataset itself, and we may have anisotropy, so the distance $\mathbf{x}_\alpha - \mathbf{x}$ is no longer as meaningful.
- RBF interpolation only returns an interpolation, it does not provide a measure of confidence in that interpolation, such as the kriging variance.

RBF interpolation works well when you have a lot of data, or a lot of noisy data, preferably on a regular grid, and the purpose is to remove the noise.

 Visit **Notebook 17: Radial Basis Function (RBF)** to experiment with the Radial Basis Function (RBF) interpolation and smoothing and creating the trend model of Figure 4.28.

There are quantitative rules that tell us how much variation we should put in the trend $m(\mathbf{x})$ and how much should be put in the residual $R(\mathbf{x})$. A few guiding principles are:

- The residual should no longer show trend; this can easily be checked by calculating the variogram of the residual data and checking if it has a constant sill in all directions.
- Since the trend function is deterministic, we should have high confidence in that function. This can be tested by removing some data and seeing how much the function fluctuates.

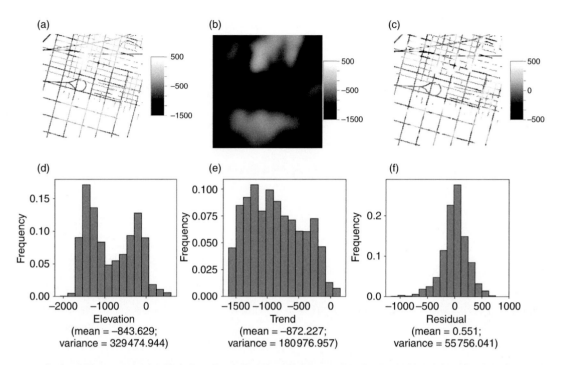

Figure 4.29 (a) Original flightline data. (b) The trend is modeled using RBF smoothing. (c) The residual obtained after subtracting the trend. (d)–(f) Histograms of the elevation line data, trend, and residual, respectively.

Figure 4.29 is important in showing the variance of each component, as displayed in the histograms. The largest portion of variance is explained by the trend (~180 000), the residual has much less variance (~55 000).

Locally Varying Mean Kriging

The trend function obtained in the previous section can be used in several kriging methods that account for this trend. While many techniques exist (kriging with trend, kriging with external drift), we will limit ourselves to the easiest to apply, kriging with a locally varying mean. In this form of kriging, we split the problem of kriging into two separate tasks using the protocol in Box 4.2.

Because kriging is an exact interpolator, the residual at the data location is $r(\mathbf{x}_\alpha)$; hence, adding the residual to the trend function gives back exactly the original data $z(\mathbf{x}_\alpha)$.

4.5.6   Worked-Out Example: Kriging the Thwaites glaicer in Antarctica

We are now ready to apply kriging with locally varying mean to the flightline data.

Box 4.2 Protocol for kriging with a locally varying mean

- Perform a trend function analysis to get $m(\mathbf{x})$
- Subtract the trend function $m(\mathbf{x})$ from the data $z(\mathbf{x}_\alpha)$ to get the residual data $r(\mathbf{x}_\alpha)$
- Model the variogram of the residual using the variogram modeling protocol
- Perform kriging of the residual using that variogram
- Add the kriging map of the residual to the trend function $m(\mathbf{x})$

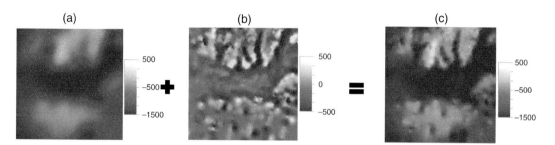

Figure 4.30 We add (a) the trend model to (b) the kriged residual to create (c) the final kriging with a locally varying mean result.

- Carry out a trend function analysis, as detailed in the section above.
- Subtract the trend to get residual data; see Figure 4.29c.
- Calculate the variogram and apply variogram interpretation protocol; see the section above, "A variogram interpretation protocol for two dimensions."
- Apply kriging of the residual: here we use local ordinary kriging, as detailed in the section above.
- Add the trend to the interpolated residual; see Figure 4.30.

4.5.7 What Have We Learned about Kriging?

- We have learned that kriging is a spatial interpolator relying on two principles: (1) it accounts for spatial correlation between the data and the unknown true value; (2) it accounts for the spatial correlation between the data.
- We have learned that kriging in practice requires specifying a search neighborhood and that some attention to the specification of this neighborhood is required to avoid artifacts in the kriging maps.
- We have learned that kriging in the presence of a trend requires removing the trend. The resulting residual variogram is then used for kriging.
- We have learned about the kriging variance, which is an estimation variance. It quantifies the confidence in the kriging estimate.

4.6   Conditional Simulation

Kriging is but one of many possible interpolation methods. Other methods are spline interpolation or the RBF interpolation considered above. Regardless of what interpolation method you use, there is one common truth: the maps you create are not real! We want a method that can generate more realistic maps. If possible, we want to general many different realistic maps so that we can get close to the true values of the flightline gaps.

4.6.1   The Smoothing Property Common to All Interpolators

What do we mean by "not real"? Just consider one simple statistical summary: we calculate the variance of the data and the variance of the interpolations made; see Figure 4.31. Clearly the data have more variance. Also, we can see that the two peaks in this histogram are different. The kriging interpolators smooth out the real variability between valley and mountain. So, our interpolated map is not real in terms of variance.

Why is the happening? We need to realize that all interpolation methods have two goals in mind: (1) to match the data and (2) to be as close as possible to the unknown truth. The latter is what is causing the problem. Let's focus on a very simple analogy. Consider a coin toss where we assign the value of 1 to heads and 0 to tails. Now, you need to make a call, but your task is to be as close as possible to the unknown truth (1 or 0). You would guess 0.5, a value that is not even real! It is the value that is on average closest to 0 and 1. Interpolations therefore create unrealistic values; the map you are looking at is very deceiving in that sense. Unfortunately, it is quite common practice: geoscientists interpolate data, then look at a map as if it is real.

The quest therefore is to create maps that look real, in the sense that they have variance (and variograms) similar to the data. The other side of this coin then is, as we will see, that we no longer have a unique map, we are no longer looking for that single answer that is as close as possible to the truth. Instead, we generate many maps, which we term "realizations"; we can

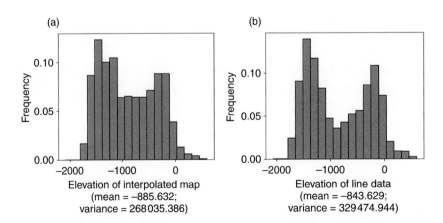

Figure 4.31 Comparing histograms of (a) the interpolated map and (b) the data used for interpolation. Interpolation is a smoother method because the variance of the map is less than the variance of the data.

even make a movie of these maps, to show with a single image (the "movie") where we have most uncertainty. This concept is termed spatial uncertainty.

Interpolated maps can be used as a visual summary of the spatial data at hand. However, that would only be of very limited use. Interpolated data may be used for many purposes:

- as data for further spatial analysis and data aggregation, see for example in Chapter 3
- for physical forward modeling, such as groundwaterflow modeling

4.6.2 Sequential Simulation

The kriging maps are missing variance relative to the data used in kriging. To restore this variance, we rely on the estimation variance, or kriging variance. Instead of having a single value at that location, we now obtain an uncertainty statement, called local uncertainty. We then use this local uncertainty model to restore the variance by creating multiple maps that have the same variance and variograms as the data.

Local Uncertainty

Having identified the smoothing property of all interpolators, and the biases this may incur when using these maps in further analysis, we need to come up with alternative ways of interpolating. Let's revisit the kriging estimator:

$$\hat{z}(\mathbf{x}) = \sum_{\alpha=1}^{n} \lambda_\alpha(\mathbf{x}) z(\mathbf{x}_\alpha) \tag{4.21}$$

We now know that the variance of $\hat{z}(\mathbf{x})$ is less than the variance of the data $z(\mathbf{x}_\alpha)$. How much variance are we lacking? Geostatisticians have found that that exact amount is equal to the kriging variance:

$$var(\mathbf{x}) = var(z) - \sum_{\alpha=1}^{n} \lambda_\alpha \Big(var(z) - \gamma(\mathbf{h}_{0\alpha}) \Big) \tag{4.22}$$

Let's look at some limiting cases. At the data locations, the kriging variance is exactly zero. Kriging is an exact interpolator; it matches the data at the data locations. Hence, at those locations we are not lacking any variance. If we go progressively away from the data, then the estimator will become equal to the mean, a constant, at least when performing simple kriging. In that case, the variance we lack is the variance of the data. The way in which we go from there being no variance missing to all variance missing has to be the function of the variogram. The shorter the range of the variogram, the quicker this missing variance increases. The above equation expresses that functionality.

The next question is how to restore the variance. To do that, we will start, for now, with the assumption that data exhibit a standard Gaussian distribution. Instead of assigning a single estimate $\hat{z}(\mathbf{x})$ at location $\mathbf{x}$, we assign a probability distribution at that location, namely a Gaussian distribution whose mean is equal to $\hat{z}(\mathbf{x})$ and whose variance is equal to $var(\mathbf{x})$. Given the discussion above, this makes logical sense: we are adding variance by assuming a probability distribution instead of a single estimate. What does this mean practically? There are two such uses. The first is that at each location we now have an uncertainty quantification in the form of a Gaussian distribution; hence, we can for example predict probabilities over some specified threshold. The second one is that we can sample from this distribution using Monte Carlo simulation (see Section 5.2.3). The latter means that at that location we no longer assign a single value, but a number of alternative values, sampled from the Gaussian distribution. We term this type of uncertainty as local uncertainty. The local refers to it being assigned to a specific location.

We take the concept of Monte Carlo simulation from a local probability distribution further, assume a standard Gaussian distribution, and create maps whose variance is equal to the variance of the standard Gaussian distribution, namely "1."

Sequential Simulation

We continue working under the assumption that the data are standard Gaussian. Imagine now that we had picked, randomly, any location in space that is not a data location; see Figure 4.32. At that location, we sample a single value from the local probability distribution. Let's term the simulated value $z^s(\mathbf{x}_0)$. Let's now, again randomly, select another location. What is the local probability distribution now? We repeat the same kriging procedure

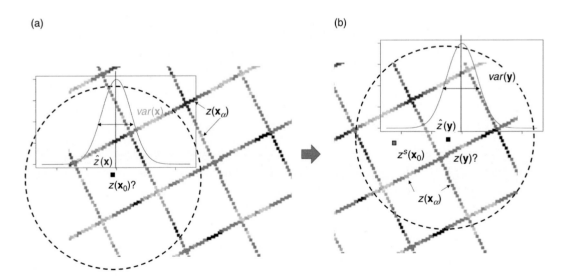

Figure 4.32 The idea of sequential simulation is (a) to draw simulated values from a probability density function, instead of assigning the kriging estimate to the unsampled location; then (b) at the next randomly picked location we use the previously simulated value as data.

with one important difference: in kriging we also take the previously sampled value as if it were real data:

$$\hat{z}(\mathbf{y}) = \sum_{\alpha'=1}^{n} \lambda_{\alpha'}(\mathbf{y}) z(\mathbf{x}_{\alpha'}) + \lambda_{n+1} z^s(\mathbf{x}_0) \tag{4.23}$$

And hence the simple kriging variance is:

$$var(\mathbf{y}) = var(z) - \left(\sum_{\alpha'=1}^{n} \lambda_{\alpha'} \left(var(z) - \gamma\left(\mathbf{h}_{y\alpha'}\right) \right) + \lambda_{n+1} \left(var(z) - \gamma(\mathbf{h}_{y0}) \right) \right) \tag{4.24}$$

We sample from a Gaussian distribution with mean $\hat{z}(\mathbf{y})$ and variance $var(\mathbf{y})$, then assign that sampled value to location $\mathbf{y}$. We visit each unsampled location in the grid in a sequential random order until all values are simulated, however, at each location, accounting for the previously simulated values. We term this form of simulation sequential simulation, because of the sequential visit in random order of each grid location. What have we achieved? We have created a map with two important statistical properties:

- the histogram of this map is standard Gaussian
- the variogram equals the variogram derived from the data using the variogram modeling protocol

We have also achieved a third feat. We can create many maps this way, each time visiting the locations in the grid in a different random order. The order of visits is also termed the random path.

Constraining to the Histogram

The concepts developed so far rely on the assumption that the data have a standard Gaussian distribution. This is rarely the case in reality; hence, we need to address this situation. A common practice in data science is to perform a transformation of the data such that the data become standard Gaussian. The tool to do this is called "normal-score transformation." In fact, one can always transform, bijectivity, any set of continuous values into any other set, using a monotonic, rank-preserving transformation, not just into a standard Gaussian. The idea is described in Figure 4.33. Any value in the original dataset can be related to a quantile. For example, the middle value is the 50% quantile. For the standard normal distribution, we know the 50% quantile, it equals 0. Hence, the median in our dataset is always transformed to zero. Because each sample value in our dataset is a quantile, we can repeat this for all samples to find that the dataset is not standard Gaussian. You notice in Figure 4.33 that this operation is completely reversible. Hence, once we have done our statistical work on the standard Gaussian values, we can back-transform them into the original values.

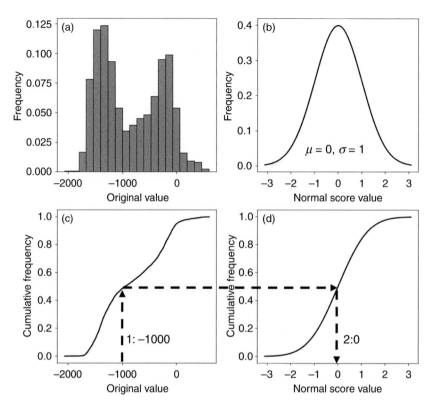

Figure 4.33 Example of the normal-score transform and back-transform. (a) Histogram of the original data values that do not distribute as Gaussian. (b) Standard Gaussian distribution. (c) Transformation of any original value (example −1000) using the original data cumulative frequency to (d) a normal score value, corresponding to the same cumulative frequency of standard Gaussian distribution (the example reads as 0).

Sequential Gaussian Simulation Protocol

Because sequential Gaussian simulation recycles a lot of ideas from kriging, we can keep many of the protocols, such as the variogram interpretation protocol. Box 4.3 gives the protocol for performing sequential Gaussian simulation:

 Play **Video 10: Simulation** to simulate spatial data using sequential Gaussian simulation. This includes the tutorial on using SGEMS for simulation.

4.6.3 Application to the Danish Case

In this section, we work out in detail the steps needed to perform a sequential simulation and use the results to make predictions of redox conditions. Here are the steps to make such prediction:

1. Generate many three-dimensional realizations of resistivity using the tTEM soundings of Figure 4.4.

Box 4.3 Sequential Gaussian simulation protocol

- Perform normal-score transformation of the data such that they are standard Gaussian.
- Check whether the normal-score transform worked by making a normal-quantile plot
- Apply the variogram calculation and modeling protocol to the standard Gaussian data
- Create a grid on which to perform the simulations
- Specify the number of simulations you want to create
- Specify the search neighborhood (as was done in kriging)
- Execute sequential Gaussian simulation
- Quality control 1: make a quantile–quantile plot of a simulated realization versus the data
- Quality control 2: calculate the variogram of the simulated realization and compare with the variogram of the data. This can be done with or without normal-score transformation.
- Make a few summary statistics based on all realizations

Plots to make
- Normal quantile plot to check the normal-score transform
- Plots for making variograms following the protocol in Box 4.1
- Quantile–quantile plot of a simulated realization versus the data
- Variogram of simulated realizations versus variogram the data

---

2. Use each realization as a predictor to perform logistic regression (Chapter 3) with the borehole data.
3. Aggregate the predictions made based on these multiple realizations into one single prediction. Here we will use the concept of log ratios, which were discussed in Chapter 2.

 Visit **Notebook 18: Danish Case** to follow the tTEM simulation and redox prediction steps.

The Danish case is a three-dimensional case, so we'll show here how that is different from two dimensions. In fact, it is not very different. In three dimensions we have an ellipsoid, which is defined using three axes and three angles. In most practical cases, certainly in the subsurface, you won't need six values. We only need four values, three axes and one angle: the azimuth angle. Why is this? Subsurface geological variability is often layered, so usually two distinct variations exist: one in the vertical, and one in the horizontal direction. Typically, along the vertical direction you will find much more variability compared to the horizontal.

In the Danish case, a vertical trend exists, as shown in see Figure 4.34a; hence, as before, we first perform a trend function analysis (e.g., using RBF smoothing), then subtract the trend from the data. After trend removal, we calculate the residual variograms; see Figure 4.34b. We apply the variogram interpolation protocol in three dimensions to get the variogram parameters required for performing sequential Gaussian simulation. Four elements are required for sequential Gaussian simulation:

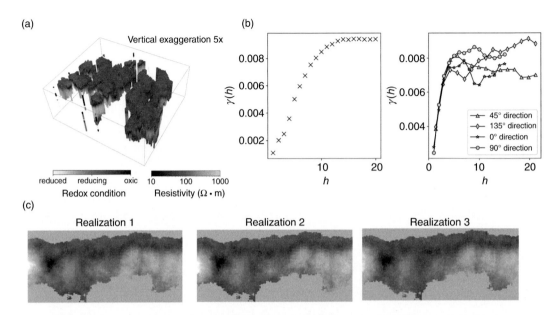

Figure 4.34 (a) Non-collocated borehole and transient electromagnetic data. (b) Variograms in vertical (resolution 1 m) and horizontal directions (resolution 20 m). (c) Three out of 100 realizations generated using sequential Gaussian simulation.

- the three-dimensional grid on which to simulate values
- the data used (here the various one-dimensional electromagnetic measurements)
- the histogram of the electromagnetic data
- the variogram of the normal-score data

To go from resistivity to denitrification potential we use logistic regression (see Chapter 3). The input to logistic regression is the resistivity in a neighborhood area around the borehole, as well as the depth; see Figure 4.35a. We get as many logistic regression results as we have realizations of the electromagnetic data. All 100 logistic regression probabilities are aggerated into a single posterior probability (Figure 4.35c), including a measure of confidence (Figure 4.35e). Figure 4.35d shows the final classification result that can be used to inform decisions related to sustainable farming.

4.6.4   What Have We Learned about Conditional Simulation?

- We have learned that conditional simulation creates spatial maps with the same variance and variograms as the data
- We have learned that, in practice, conditional simulation also needs trend removal, which is similar to kriging.

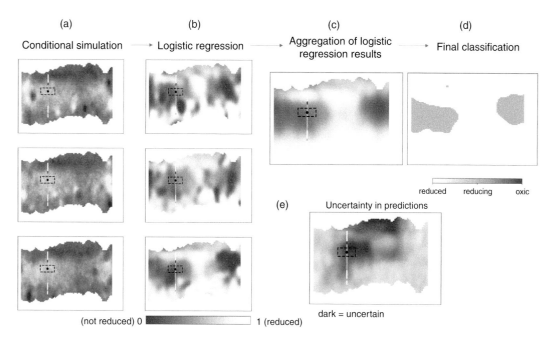

Figure 4.35 (a) Conditional simulation: three realizations (out of 100) of the electromagnetic data juxtaposed to a single borehole; the dashed-line box indicates the local electromagnetic data used to perform logistic regression. (b) Resulting logistic regression applied to each realization. Results show the probability of being reduced. (c) Aggregation of all logistic regression results in a single posterior probability. (d) Final classification result. (e) Measure of confidence in the classification.

4.7 Multiple-Point Geostatistics

Multiple-point geostatistics (MPS) is the field of study that focuses on the digital representation of physical reality by reproducing high-order statistics inferred from training data, usually training images, that represent the spatial (and temporal) patterns expected in such contexts. To understand the usefulness of an approach different from variograms, we need to consider the limitations of variogram-based methods.

4.7.1 Limitations of Variograms

Consider an unknown true topography in Figure 4.36a. Instead, what are available are limited flightline data (Figure 4.36b). The approach we have taken so far is to estimate the variogram from the data (Figure 4.36c, where the unit for h is 500 m), then use this variogram in kriging or conditional simulation. The simulated realizations shown in Figure 4.36d are reflective of the truth in the sense that they share the same variogram and histogram. Yet, when looking at these simulated realizations, we can see there are two important differences between the simulations and the truth:

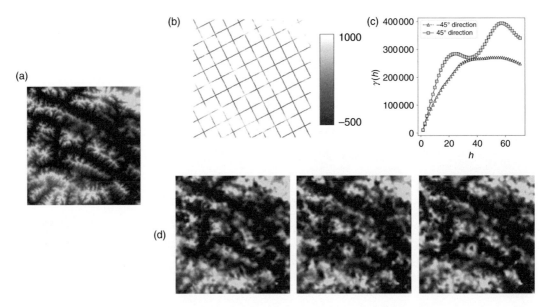

Figure 4.36 Multiple realizations of topography generated using sequential Gaussian variation. See text for details.

- the strong connectivity of the glaciated valleys that really exists is not present in the realizations
- the smaller-scale variations of incised mountain ridges are not present in the simulations

Clearly, the variogram is limited in capturing spatial variability. The approach taken in this section is to drop variograms altogether. Instead, we convey spatial variability through an example termed a "training image." Figure 4.37 shows the approach we develop in this section. Instead of calculating variograms from data, we provide an image that we deem similar in variation to the unknown truth (Figure 4.37a). In the Antarctica case, we hypothesize that the Arctic is representative for Antarctica; hence, we use a portion of the Arctic topography as the training image (Figure 4.37c). Then, conditional simulation algorithms aim to capture the complex variation in the training image and generate realizations constrained to the flightline data, as shown in Figure 4.37. We have now introduced two new elements that we discuss next: (1) training images and (2) the algorithms that achieve the stated aims.

4.7.2   What Is a Training Image?

Training images are datasets that are exhaustive in the sense that they are a single truth fully describing two-, three-, or even four-dimensional (space–time) variation. In that sense, the name "image" is a bit deceiving because that image may be four dimensional. The data conveyed in training images cover an entire spatial or spatio-temporal domain. As a result, much more complex variation can be captured compared to variograms. As mentioned above, MPS methodology involves the use of training images. The variogram can be seen as a two-point

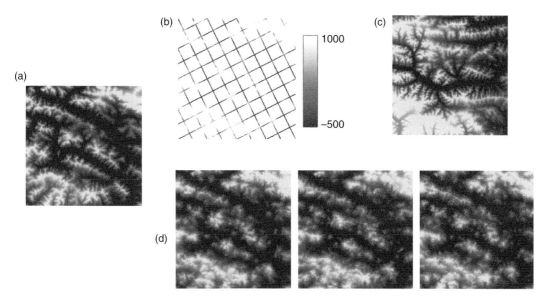

Figure 4.37 In MPS we use a training image instead of a variogram to generate realization constrained to the flightline data: (a) unknown truth; (b) flightline data of the unknown truth; (c) training image; and (d) three realizations constrained to the flightline data.

statistical measure: it uses two values at a time (pairs of values separated by a vector h) in its calculation. In MPS, we use higher-order or multiple-point statistical features.

A key requirement for the training image is for it to be representative of the domain to be simulated. This requirement is difficult to assess; a necessary component is that the training image is required to have the same histogram and variogram of the data. But this is not a sufficient requirement. Consider two opposite situations:

• We have very few data: in that case, it is easy to pick a training image that is representative of the data. Also, with few data, the variogram becomes unreliable, and hence the training-image approach becomes attractive. However, many training images may now be deemed representative, and hence the uncertainty of the training-image choice becomes important.

• We have a lot of data: in that case, we are less concerned with uncertainty in selecting a training image; instead we are concerned about whether the training image is in conflict with the data.

The problem of selecting a suitable training dataset is not limited to MPS. In fact, it is a significant challenge in machine learning in general.

The other component to MPS is to develop algorithms that use training images and generate conditional realizations. Many such algorithms exist, and here we present one such algorithm: direct sampling. It is one of the easiest to use and explain, and hence it may be a good introduction for you in the area of MPS.

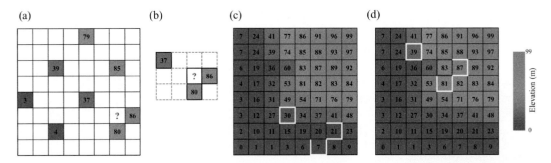

Figure 4.38 A conceptual example of DS simulation. (a) Radar lines on the simulation grid. (b) Three known points (values of 37, 80, 86) constitute a conditioning data pattern. (c) A mismatch pattern in the training image. (d) A similar pattern in the training image. (Figure from Yin et al., 2022, distributed under a CC BY 4.0 license.)

4.7.3   Direct Sampling

Direct sampling (DS) is a widely used MPS approach for achieving spatial modeling and gap filling. Figure 4.38 provides a simplified example of DS in the context of flightlines. The values in the grid indicate the elevation. DS, like sequential Gaussian simulation, performs simulation sequentially. The DS algorithm visits an unsampled location in the simulation grid and collects neighboring data. For example, in Figure 4.38a, three conditioning points are located near an unsampled location (marked with "?"). DS records the values and relative locations of known data values. The latter is called the data event. Then, the training image is searched for similar data events. The similarity is defined by a certain distance metric (e.g., the Hamming distance for categorical variables and Euclidean distance for continuous variables). As Figure 4.38d shows, the first similar data event is located, then its central value is pasted into the simulation grid. The center of the similar instance is pasted into the simulation grid. The preceding simulation program is repeatedly performed until there is no unknown point in the grid.

We now arrive at the main input parameters required for DS, which are substantially fewer than variogram-based method:

- The amount of data in the search neighborhood n: The idea here is the same as in kriging or sequential Gaussian simulation. The larger the search neighborhood, the more time DS will require to find a similar data event in the training image.
- The distance threshold t: This threshold quantifies when a data event in the training image is deemed similar. A low value of t again will result in increased computing time, but also increased quality in the simulation in terms of matching the pattern of the training image.
- The fraction (f) of the training images that will be scanned. We may not need to scan the entire number of images; we could stop at some fraction, for example 50%.

4.7.4   Application of DS to the Antarctica Dataset

 Visit **Notebook 19: Direct Sampling (DS)** to apply DS to the Antarctica case. Test out various combinations of the tuning parameters n, t, and f, to test the sensitivity of these parameters on the result in Figure 4.38.

4.7.5   What Have We Learned about MPS?

- We have learned that MPS uses training images, instead of variograms, to quantify spatial variations.
- We have learned that MPS relies on realistic training images, which is a prerequisite of using this method: we need to prepare training images from geological knowledge or previous case studies.
- We have learned that we can scan a sample of our training images; scanning the entire number of training images to find a matching pattern might take a long time at each step.

4.8   What Have We Learned in This Chapter?

- We have learned that spatial-temporal interpolation requires quantification of spatio-temporal variability, and that the variogram is one way to achieve this.
- We have learned about a geostatistical practice that splits the problem of interpolation into two parts: estimating the trend and estimating the variation around that trend.
- We have learned that all interpolators create a map that is smoother than the actual data.
- We have learned about conditional simulation, which creates maps with the same variance and variogram as the data.
- We have learned about the limitations of the variogram in quantifying complex spatial variation and how a methodology termed multiple-point geostatistics may provide an alternative through the specification of training images.

TEST YOUR KNOWLEDGE

4.1   What is geostatistics about?
   a. Modeling spatio-temporal behavior
   b. Using the modeled spatio-temporal behavior to make a prediction at unsampled location in space–time
   c. Quantifying spatial uncertainty
   d. All of the above

4.2 What is geometric anisotropy?
  a. Variograms along different directions have a different nugget effect
  b. Variograms along different directions have a different sill
  c. Variograms along different directions have different range
4.3 What is true?
  a. All one-dimensional variograms start at (0, 0)
  b. Variograms start at (0, 0) except when there is a nugget
  c. Variograms start at (0, 0) except when there is trend
4.4 What is the minimum number of parameters required for modeling geometric anisotropy in two dimensions?
  a. Three
  b. Six
  c. Nine
4.5 What is special about spatial?
  a. Data in space always have a volume attached to them
  b. The fabric of space can be divided into infinitely small parts
  c. Space does not require a grid
  d. All of the above
4.6 What does a hole effect indicate?
  a. Important short-range fluctuations
  b. Systematic long-range fluctuations
  c. Non-stationary variance
4.7 What is the minimum number of parameters required for modeling geometric anisotropy in three dimensions?
  a. Six
  b. Eight
  c. Nine
4.8 What is the most important in kriging?
  a. To be accurate about the variogram model for the first few lag distances
  b. To be accurate about the sill
  c. To be accurate about the long-range fluctuations in the variogram

FURTHER READING

There are many great textbooks in the geostatistical community. We recommend the following three:

- Isaaks, E. H. and Srivastava, R. M. (1989). *Applied Geostatistics*. Oxford University Press.
- Remy, N., Boucher, A., and Wu, J. (2009). *Applied Geostatistics with SGeMS: A User's Guide*. Cambridge University Press.
- Mariethoz, G. and Caers, J. (2014). *Multiple-Point Geostatistics: Stochastic Modeling with Training Images*. John Wiley & Sons.

REFERENCES

Cressie, N. (1990). The origins of kriging. *Mathematical Geology*, 22(3), 239–252.

Krige, D. G. (1952). A statistical analysis of some of the borehole values in the Orange Free State goldfield. *Journal of the Southern African Institute of Mining and Metallurgy*, 53(3), 47–64.

Matheron, G. (1969). *Le krigeage universel [Universal Kriging]*. Cahiers du Centre de Morphologie Mathematique, Ecole des Mines de Paris.

Matheron, G. (1971). *La théorie des fonctions aléatoires intrinsèques généralisées*. Note du Centre de Géostatistique, Ecole des Mines de Paris.

Neven, A., Maurya, P. K., Christiansen, A. V., and Renard, P. (2021). tTEM20AAR: a benchmark geophysical data set for unconsolidated fluvioglacial sediments. *Earth System Science Data*, 13(6), 2743–2752.

Tian, H., Xu, R., Canadell, J. G., et al. (2020). A comprehensive quantification of global nitrous oxide sources and sinks. *Nature*, 586(7828), 248–256.

Yin, Z., Zuo, C., MacKie, E. J., and Caers, J. (2022). Mapping high-resolution basal topography of West Antarctica from radar data using non-stationary multiple-point geostatistics (MPS-BedMappingV1). *Geoscientific Model Development*, 15(4), 1477–1497.

5 Review of Mathematical and Statistical Concepts

5.1 Purpose of This Chapter

Previous chapters covered four data science topics relevant to the geosciences. The purpose of this chapter is different. The material is supplementary to the four major topics. The purpose is to introduce basic concepts in statistics, as they are relevant to each chapter. We are not aiming for a comprehensive, deep-reaching technical review, starting with definitions of random variables and going on to discuss distribution types, etc. Instead, we present, in line with the philosophy of this book, an intuitive, yet rigorous, introduction to many concepts. The outline follows what is relevant for each chapter, so instead of starting with a long introduction chapter, we provide this as a separate dedicated chapter, with topics developing in the same order as the preceding chapters. This allows students and teacher to review the matter as is relevant and in connection with the actual real-world examples always in mind.

This chapter doesn't include questions, notebooks, or exercises. Instead, we recommend, among the many other excellent texts, the following data science and machine-learning books. These books provide a comprehensive review of probability, statistics, and machine-learning methods with many general scientific applications. We recommend that students first read this chapter for a review of statistical concepts and refer to the recommended books as needed:

1. You can delve deeper into probability and statistics, which considers both frequentist and Bayesian aspects, in *Probability and Statistics*, by M. H. DeGroot and M. J. Schervish (Pearson Education, 2012).
2. You will find a more comprehensive review of the bootstrap method and a history of statistics over the last 60 years in *Computer Age Statistical Inference, Student Edition: Algorithms, Evidence, and Data Science*, by B. Efron and T. Hastie (Cambridge University Press, 2021).
3. The book *Machine Learning*, by Z. H. Zhou (Springer, 2021), offers an introduction to popular machine learning methods, which is a good supplement to our Chapter 3.
4. You can find more instructions on how to play with your data using Python in *Data Science From Scratch: First Principles with Python*, by J. Grus (O'Reilly Media, 2019).

5.2   Concept Review for Chapter 1: Extreme Value Analysis

In Chapter 1, we cover basic distribution models, then develop the family of extreme value distributions. In this review section, we introduce summary statistics, basic distribution models, and the use of computer simulations in statistical analysis.

5.2.1   Quantiles and Percentiles

Quantiles are an important concept in data science. The empirical (arithmetic) mean and variance are very sensitive to large values; a small change in a large value will likely affect the variance considerably. This is an undesirable feature. Quantiles of data are like the mean and the variance, which are summary statistics. In defining and calculating quantiles, we need to rank data first. Consider a dataset of diamonds; we will rank them from small to large. After ranking, we use the $\leq$ symbol, meaning "less than or equal to." The "equal to" is important since we may have ties in the dataset: more than one sample with the same value. Ties are annoying, since they suggest a form of discreteness in what is considered to be a "continuous" variable (diamond size). Often in data science discrete and continuous variables are treated differently, but with ties we may have a dataset that is mixed in that sense. Some methods require that ties be "broken," in other words, even though the value is the same, you still need to come up with a rank. A sample or dataset is denoted as:

$$x_1, x_2, x_3, \ldots, \tag{5.1}$$

The "1" here indicates the first sample, not necessarily the smallest one. For ranked data, we use a special notation with * and so we use the following notation:

$$x_1^* \leq x_2^* \leq x_3^* \leq \ldots \tag{5.2}$$

This may look a bit confusing because we still have a subscript "1," but now the one is the smallest sample, not the first sample. We can also decide to rank from highest to lowest; then we write:

$$x_1^* > x_2^* > x_3^* > \ldots \tag{5.3}$$

Having defined ranks, we can now define quantiles. Let's start with the simplest: the 50% quantile, also better known as the median. To find the median, we need to count, meaning we find that value in our rank such that half the data (50% or 50 percentile) is less than or equal to that value. Consider four samples, then this will be the second-lowest value. For example:

$$1\ 3\ 8\ 9$$

Here, 3 is the value such that 50% of the values are less than or equal to 3. Now you will understand the importance of the "equal to" part. Indeed 1 and 3, two values, are less than or equal to 3. The sample values 8 and 9 are strictly larger than 3. What happens when we have an odd number of sample:

$$1\ 3\ 6\ 8\ 9$$

The middle value is "6", but 6 is associated with 60%, because three out of five are less than or equal to 6, hence the "true" median will be in between. There are a number of ways around this, one is to average 3 and 6. This is less desirable if you don't want to make any assumptions, that is, deviate from the dataset at hand. The other choice is to find the first value in the dataset such that at least 50% of the data are less or equal than that value. In this case, this is the value 6. When we have large datasets, it does not matter that much what exactly you do; however, it is preferable to define "empirical quantiles" as actual sample values (such as 6), and not something calculated from sample values (averaging 3 and 6).

So far, we have covered empirical quantiles, namely those calculated from data; next we can also define model quantiles, those calculated from a theoretical model we define. To do so we first introduction distribution models for data.

5.2.2 Distribution Models

In the previous section we discussed empirical statistical summaries. In this section, we look at ways of defining models that can describe statistical variation theoretically.

Random Variables and Outcomes

An important goal common to all applications in this book is to predict variables that are yet unknown. In mathematics, "x" is used generally for something unknown. In data science, however, we have a double meaning: x is unknown/unspecified, but x is also uncertain. Something can be unknown/unspecified, but deterministic. For example, we know it is a single number or value, we just don't know which one. In data science, we may know something about x, but in a probabilistic sense. It is important to distinguish between an unspecified outcome and an uncertain variable. In data science, when a variable is uncertain, we term it a *random variable* and we use the notation of a capital X. An unknown, but deterministic value is written with the lower-case x, which is termed the *outcome*. For this reason, we can write the following

$$P(X \leq x) \qquad (5.4)$$

The first X refers to the variable we are studying, the second x, refers to some possible value of the uncertain variable; x can be 3, for example, $P(X \leq 3)$, Therefore we can write the following

$$F(x) = P(X \leq x) \qquad (5.5)$$

We now use the function notation, because the right-hand side is a function of lower-case x. Some data scientists like to make it even more clear and write

$$F_X(x) = P(X \leq x) \qquad (5.6)$$

referring to the fact if we are dealing with a random variable X. You can also write

$$F_Y(x) = P(Y \leq x) \qquad (5.7)$$

This is a different function, because we are dealing with a different random variable Y. However, the unspecifed x (also termed threshold here) is the same in both functions.

Probability Density Functions

"Distribution" in data science is a very commonly used word. Other ways in which you will encounter this is:

- the distribution of the data is skew
- the distribution model is lognormal
- these two datasets have different distributions

Distribution comes from the Latin *dis* (apart) *tribuere* (assign). Perhaps a good analogy is how one distributes seeds in soil when gardening. The seeds are spread over some area. The spread can be uniform, or in clusters. Sample values are like these seeds, but now we start out in one dimension, namely we spread them over a real-valued axis, as shown in Figure 5.1. We notice in this case that sample values seems to be distributed into two clusters, one around 0, another around 1. Having points on an axis is not a very attractive way to represent data, hence the use of the histogram. To construct a histogram, we first create bins (Figure 5.2a and 5.2b) and then count how many samples there are in each bin for a given bin or window size. However, we can make a histogram another way, not by binning but by first selecting a window size (0.2) and sliding a window over the axis, each time with a smaller slide (0.05) compared to the window size (Figure 5.2c). This creates a much smoother-looking plot than the histogram plot. Another

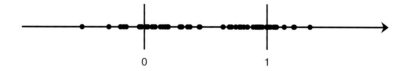

Figure 5.1 Samples distributed align on an axis.

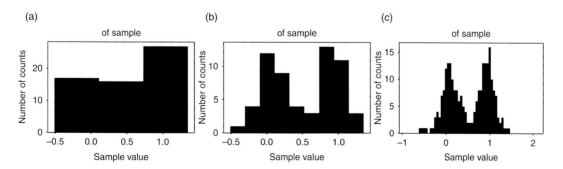

Figure 5.2 Histograms with window sizes of (a) 0.5 and (b) 0.2. (c) Histogram created using a sliding window with the window = 0.2, size of each slide = 0.05.

approach is to put a little window over each data point and add these small windows into one single plot. All these approaches are a function of the size of the window you use.

The values created this way are "numbers of sample values per windowed interval"; this is also a density value. How many seeds are there per m^2? How much mass is there in a volume? All these are questions about density. A density is a property that is a ratio, hence we no longer need to worry about "volume" or "window size." Density works for very small as well as for very large volumes. So, the curve is in that way a representation of how the density of a sample values varies along the axis.

To understand the nature of theoretical models in data science, we encourage you to think about the situation of "having a very large number of samples." In such cases, we can make the window very small and still get nice smooth results. That also means that, if we have an infinite number of samples, we can make the window infinitesimally small. If you believe this is strange and confusing, then indeed, what is being done here is purely mathematical, and only for reasons of convenience. In reality, there will never be infinitely large amounts of data in combination with infinitesimally small windows. Still, it is useful to think that way simply because then we can get rid of the influence of the window size when we define the mathematical model.

Let's now investigate what happens when we go from a finite sample dataset to a very large (possibly infinite) dataset, the latter we then call a "model." We will start with a finite sample and count how many samples fall in an interval starting at x_i and window size Δx, and divide this count by the total sample size n to get a frequency

$$\text{Frequency}(x_i < x < x_i + \Delta x) = \frac{n_i}{n} \tag{5.8}$$

This value depends not only on Δx but also on the "density" of samples, let's term this density

$$f_i = \frac{n_i}{\Delta x^* n} \tag{5.9}$$

and

$$f_j = \frac{n_j}{\Delta x^* n} \tag{5.10}$$

where f is a density in the sense that we divide by some length, namely Δx. A very simple way to get rid of the window and the sample size n is to takethe ratio

$$f_i/f_j = n_i/n_j \tag{5.11}$$

This ratio expresses how much more likely one will get samples near x_i vs x_j. Indeed, if for example that ratio is 4, we expect four times more samples near x_i compared to x_j, yet we never need to state exactly how many, nor what the window size is!

The concept that is being introduced here is therefore not absolute but a relative comparison. One of the problems with probability is that it is difficult to make comparisons because probabilities are always between 0 and 1.

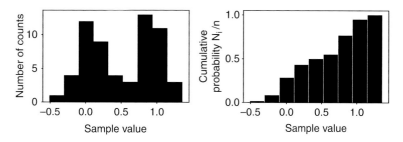

Figure 5.3 (a) Histogram with a window size of 0.2 and (b) a cumulative histogram.

The values f_i and f_j depend on where x_i and x_j are located on the x-axis, hence we can introduce the notion of a function:

$$f_i \to f(x_i); \ f_j \to f\left(x_j\right) \tag{5.12}$$

where $f(x)$ is termed the *probability density function* (PDF). We also have an extended notation: $f_X(x)$, if we are dealing with a random variable X.

We know why it is called "density," but what about the term "probability"? Density functions can be used to calculate probabilities. To see that, consider now the number of samples less or equal than x_i. Let's call this amount N_i, hence the proportion N_i/n is the frequency of samples less than or equal to x_i. In terms of the windowed histogram, see Figure 5.3b, N_i/n is calculated by summing the amounts in all windows less than x_i. This means we are summing

$$f_1 \Delta x + f_2 \Delta x + \ldots f_i \Delta x = N_i/n \tag{5.13}$$

As we take increasingly small windows, the sum becomes an integral, and the frequency becomes probability for a sample to be less than or equal to x_i:

$$x_i = \int_{-\infty}^{x_i} f(x)\mathrm{d}x \tag{5.14}$$

Cumulative Distribution Functions

There is yet another way to present data and that is via the empirical cumulative distribution function. The idea is very simple; see Figure 5.4.

- Rank a dataset with n samples from small to large.
- Plot the data by making a step of size $1/n$ on the y-axis.

Since we accumulate steps of size $1/n$, it is termed the cumulative representation of the data, or the empirical cumulative distribution (empirical = data). By doing this, we associate with each data point a value of $1/n$, and these can be interpreted in two ways:

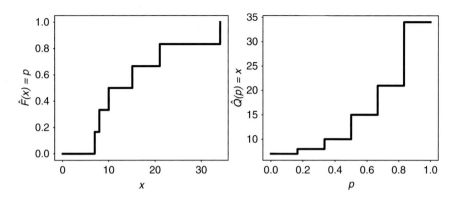

Figure 5.4 (a) Empirical cumulative distribution function $\hat{F}(x) = p$ and (b) empirical quantile function $\hat{Q}(p) = x$ of the dataset: 7,8,10,15,21,34.

- Weights: one may also choose not to use $1/n$ and weight samples differently. This weighted sampling is relevant in spatial sampling when you have clusters of data in space; hence the sampling is not uniform, but targeted. Typically, values that are clustered get less weight than values that are isolated.
- Probability: if we write the value on a ball and put these in a raffle, the probability of drawing a value is $1/n$. If we ranked the samples, the probability of drawing a sample less or equal to the i-th sample value equals i/n. That probability is the cumulative probability that we can read off the y-axis in Figure 5.4a.

We can do the same with any probability distribution function. The summing of steps of $1/n$ now becomes the integral:

$$F(x) = P(X \le x) = \int_{-\infty}^{x_i} f(x)\mathrm{d}x \tag{5.15}$$

$F(x)$ is termed the *cumulative distribution function* (CDF).

The Quantile Function

We can switch the axis in Figure 5.4a and make the y-axis the x-axis, and obtain the new staircase function; see Figure 5.4b. This inverse staircase function is termed the empirical quantile function. We denote the staircase cumulative distribution function as:

$$\hat{F}(x) = p \tag{5.16}$$

The hat means "empirical." We also write the empirical quantile function

$$\hat{Q}(p) = x \tag{5.17}$$

Likewise, if we state a cumulative distribution function as model

$$F(x) = p \tag{5.18}$$

we can derive from it a quantile function model

$$Q(p) = x = F^{-1}(p) \tag{5.19}$$

Expectation and Variance

The empirical mean and variance are summary statistics calculated from sample values. Because the number of samples is limited, the mean and variance of the entire population remain unknown. Empirical means and variances are estimates, they are very likely not equal to the population mean. What is the meaning of "population"? You can think of "population" as the complete set of samples if exhaustive sampling is performed. This requires defining what the exhaustive set is, which is a subjective definition. This is certainly true in the geosciences. For example, one may call the complete set those samples taken all over a domain in space and/or time. This requires stating the size of the domain, as well as the volume of the sampling (the sampling volume is covered in more detail in Chapter 4). Even though in reality, with a finite domain, and with a given volume, we can only sample a finite number of samples exhaustively. Still, mathematically, it is easier to work with the idea that an infinite number of samples can be obtained. In that case, the infinite number of samples, or population, can be uniquely characterized using a PDF. Then we define the theoretical population mean, also termed expectation or expected value, as follows:

$$E[X] = \int\limits_{-\infty}^{+\infty} x f(x) \mathrm{d}x \tag{5.20}$$

where $E[X]$ is the notation use for expected value of a random variable X. Where does this equation come from? Consider the equation for the empirical mean but written slightly differently:

$$\bar{x} = \sum_{i=1}^{n} \frac{1}{n} \times x_i \tag{5.21}$$

In this form, $1/n$ is a "weight" associated with sample x_i when we sample randomly. In the population, we have infinite samples, so we can no longer use $1/n$. Instead, we use the PDF to associate a weight of $f(x)\Delta x$ to an outcome x in the interval Δx: more likely outcomes should get a larger weight, and hence we get:

$$\bar{x} = \sum_{i=1}^{n} f(x_i) \Delta x \times x_i \tag{5.22}$$

Because this sum is over infinite values, we write it as an integral as in Eq. (5.20).

Likewise, we define the theoretical variance as:

$$var[X] = \int_{-\infty}^{+\infty} (x - E[X])^2 f(x) dx \tag{5.23}$$

The Exponential Distribution

Now that we are detached from any sampling, we can imagine any PDF. However, our imagining is constrained. That function needs to be positive and the integral over the range of values needs to sum/integrate to one. A very simple function has that property, namely the exponential function:

$$f(x) = \exp(-x) \tag{5.24}$$

The other nice property as you will recall from calculus courses is that the integral of the exponential function is again an exponential function:

$$F(x) = 1 - \exp(-x) \tag{5.25}$$

For this very reason, many analytical equations for density functions have some form of exp() in them. We can extend the above equations and add a parameter to the function, while still maintaining the same properties (integral and positive):

$$f(x, \lambda) = \lambda \exp(-\lambda x)$$
$$F(x, \lambda) = 1 - \exp(-\lambda x) \tag{5.26}$$

The λ regulates how fast the exponential function declines. The larger the λ, the fewer large values we expect under this model (Figure 5.5).

Kernel Density Estimation

The simplest form of density estimation is the histogram. This histogram is straightforward to compute, but it results in discontinuities in the estimated density due to the discrete nature induced by binning. Also, with histograms, no tail extrapolations are made. An other way to do this is to assume parametric functions, such as the lognormal or Pareto distribution, and fit these to the data. The main problem here is that we need to choose such a parametric form. A method that lies in between these two extremes (histogram method vs parametric form) is the kernel density estimation. The idea is explained in Figure 5.6.

In Figure 5.6 we have six samples, hence the histogram is not a reliable estimate of the density. Instead, in Figure 5.6, we put at the sample location x_i a kernel function (the dotted lines):

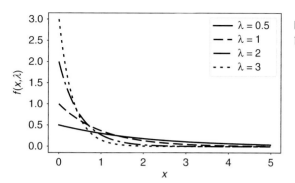

Figure 5.5 Exponential distribution with various λ.

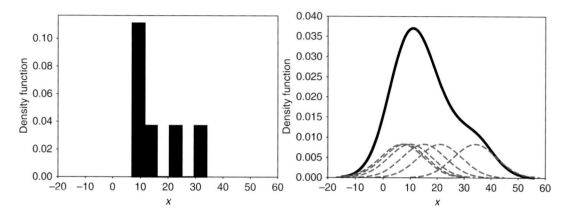

Figure 5.6 (a) Histogram of six sample values and (b) the corresponding kernel density estimation/smoothing.

$$K(x - x_i) = \frac{1}{\sqrt{2\pi}\sigma} exp\left(-\frac{(x - x_i)^2}{2\sigma^2}\right) \tag{5.27}$$

The σ determines the width of this function. The density estimate $\hat{f}(x)$ is obtained by adding up these individual kernel functions:

$$\hat{f}(x) = \frac{1}{L}\sum_{i=1}^{L} K(x - x_i) \tag{5.28}$$

where L is the number of samples (six in Figure 5.6). This equation requires the specification of one parameter: σ. Before getting to some rule of thumb for the value of σ, let's consider what σ should depend on:

- The variance of the data: if the variance is large, the data are more spread out, so we need a wider kernel, hence larger σ.
- The number of samples: If we have very few samples, there are likely to be large gaps between the data, hence the σ needs to be larger.

Using some theoretical assumptions, the following rule of thumb has been established (Silverman, 1986):

$$\sigma = \left(\frac{4}{3}\right)^{\frac{1}{3}} \times L - \frac{1}{5} \times var(\text{data}) \tag{5.29}$$

In theory, kernel density estimation can be extended to any number of dimensions. However, in practice, due to the curse of dimensionality, the number of samples required for accurate estimation grows exponentially with dimension. Silverman (1986) provides an optimistic upper bound on the number of dimensions as five.

Models Versus Reality

It is important to realize that these theoretical concepts are theoretical models that are "forced to work" in reality. In real cases, there is no hypothetical infinite sample, nor is there even a continuous real value. We represent values in a finite and discrete way, with limited number of integers behind the decimal point. That's why we call them "models," idealized versions of reality. For that very reason, there is no "true λ" in reality, because all populations contain finite samples. For example, if one would study the length of the interval between two earthquakes over the last 100 years in some area in the world, we know for sure the population has a finite number of samples. Another apparent contradiction is that for a continuous random variable, with a given PDF f, we have that

$$P(X = x) = 0 \tag{5.30}$$

This is simply due to the mathematics, nothing to do with reality, because, of course, we have some probability for X to be equal to x. What is happening here? Because probability densities are defined on the real line, this line has an infinite number of possible values, even for a finite interval, so if we calculate the probability, we get

$$P(X = x) = 1/\infty \tag{5.31}$$

For that reason, "$X = x$" is only used when X is a discrete variable; when X is continuous we use $X \leq x$ or $X > x$.

Discrete Distributions: The Poisson Distribution

So far, we have worked with values that are continuous, which means there are possibly infinite values even over a finite interval. In many cases, only a finite number of outcomes are possible. There are two cases of this nature:

- a discrete variable: the value is a number
- a categorical variable: the value is not a number, rather a non-numerical label

A specific case occurs when we count things. This case is, however, very common in the geosciences. Counting the numbers of diamonds in a sample or the number of earthquakes over some time interval. Consider the case in Figure 5.7: imagine these are sharks in some area of the ocean. We know sharks are solitary, so they don't swim together; instead we assume here that their occurrence is random. Consider now that we divide the entire area into smaller windows, and we count the number of sharks in each window, then make a histogram, see Figure 5.7. We also calculate the average for each case. This average increases per window size. The average is a number/area, so again it is a form of "density," in this setting it is also called "intensity."

Consider now the following equation:

$$P(N = n) = \frac{\exp(-\lambda \times |\text{window}|) \; (\lambda \times |\text{window}|)^n}{n!} \tag{5.32}$$

where $|\text{window}|$ is the area of the window. Now we plot this probability function on each histogram (Figure 5.7b), the empirical observed frequency vs the model. We find that the equation works for all window sizes. In fact, it will work for any shape of the window, as long as we know its area. We also notice that increasing the window size increases the symmetry of the distribution.

A French mathematician named Poisson (also French for fish) was able to show that if "fish" or any phenomenon occur randomly in space or time, the counts you make follow the above formula.

Another interesting aspect of the Poisson model is the connection with the exponential distribution. Imagine studying very large earthquakes in some region of the world, and look at the time occurrence of these earthquakes. Large earthquakes tend to be randomly occurring. Since now we are in time, and not space, we can look at the time between two earthquakes. In data science, this is termed the interarrival time. Again, Poisson showed that if events are random with intensity λ, the length of the intervals follows an exponential model.

Let's now revisit the volcano dataset and focus on the peak over threshold; see Figure 5.8. The higher we take the threshold, the more likely it becomes that the events are independent, meaning random in time. If we look at a high threshold namely $M = 5.5$, we observe, again a similar random behavior. That means that we now have a probability model for the occurrence of an extreme event. We only need to know λ.

Another way of looking at discrete events in time is to consider what the time span is between them. If a large eruption, larger than 5.5, occurs today, how long, on average, do we need to wait to see another one? This model of "waiting time" or "interarrival time" is the exponential distribution. If we call the unknown waiting time T, the probability of the interarrival time $T > t$ is:

$$P(T > t) = e^{-\lambda t} \tag{5.33}$$

Hence, by estimating λ, we can predict how long it will take for the next large event to arrive.

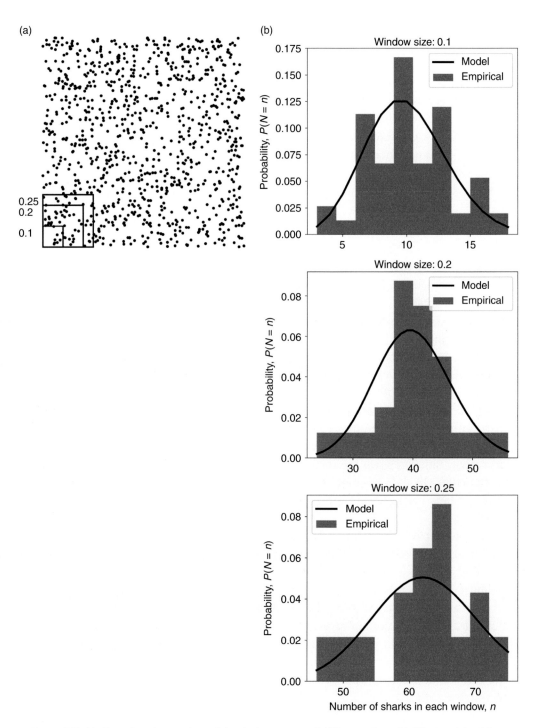

Figure 5.7 (a) Counting the number of sharks in squares of different sizes. (b) The resulting histogram overlayed with the Poisson model.

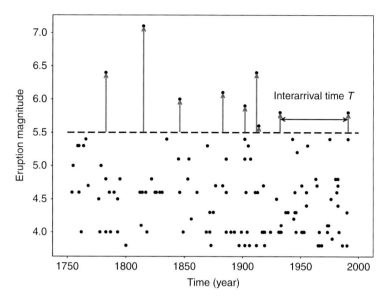

Figure 5.8 The randomness of large volcano eruptions ($M > 5.5$).

5.2.3 Sampling in the Real World and on a Computer

"Obtaining data" has many alternative wordings in data science. A very common one is "sampling," which means "getting samples." Another commonly used term that is less intuitive is "drawing." In non-science arenas, drawing is more commonly known as "sketching," which is an unfortunate common problem of English since this word also means related to sampling. A good analogy is "drawing lots" or, slightly related, "drawing funds," more fully written as "withdrawing." The best way to make the analogy is as if we were to write the value of all possible samples on a small ball and put these balls into a raffle drum. Then drawing samples is like drawing lots.

We can go out in the field and get samples or draw samples. Another use of this notion is to "draw samples" from a distribution function. In this setting, one can imagine, again, a very large number of balls in a giant raffle drum, then drawing one of them. Now sampling or drawing is theoretical: we hypothesize a distribution, an imaginary drum with a very large number of balls, and draw samples from this distribution. This notion is also termed "Monte Carlo simulation."

Monte Carlo is used in many fields of science and engineering. Broadly, Monte Carlo uses random sampling to study properties of systems with components that behave in a random fashion (Lemieux, 2009). What are these "systems," and what are such components? In our context, systems refer to physical models describing a part of the Earth. Because we do not know the Earth perfectly, the parameters of these systems remain unknown, hence are considered to be random variables. Recall that random variables are not completely unknown, often we associate probability density functions with them, much like we studied in Chapter 1. An important part of studying physical models implemented as computer programs is to observe how they behave when changing the input parameters according to their probability distributions. This requires creating samples that follow that distribution model.

In Section 1.5.2, we used sampling from distribution models to study the distribution models themselves. An equation like the lognormal, or even more complex distributions with many more parameters, may not provide much insight into what that distribution entails. Another way to study distribution models is to generate samples from them, then study the statistical behavior of these samples through summary statistics of interest, for example the frequency of exceeding a given threshold.

Next, we cover the technicalities of how Monte Carlo simulation works and generates samples from distributions.

Monte Carlo Simulation: Electronic Roulette

Writing sample values on a ball and putting them a raffle drum is fun, except it is very unpractical to sample/draw that way. We need a machine to do this for us. Our raffle drum becomes a computer program. Consider the following simple problem: we have six samples – 7, 8, 10, 15, 21, 34. We would like to write a software program that picks any value "at random" (note that "random" has a French medieval origin, "randon," meaning "rushing with disorder"). In other words, we want to remove "the order" and pick (quickly!) any.

The solution to this is not simple. In fact, the very idea was developed in the context of the Manhattan Project or "making the bomb." Much of early computing was war related. For example, the Electronic Numerical Integrator and Computer (ENIAC) was built for the computing artillery-firing tables. ENIAC weighed more than 60 000 pounds, covered 1800 square feet of area, consumed 150 kilowatts of power, and cost $500 000 to build (about $6 000 000 in today's dollars). ENIAC was the first computer on which Monte Carlo simulations took place. The main reason for using Monte Carlo simulations was not to solve data-science problems but complex physics problems, which had so many variables that the conventional deterministic technique no longer worked. Monte Carlo allowed probing solutions to complex physical problems such as the calculations for the hydrogen bomb.

John von Neumann was a Hungarian–American polymath and Manhattan Project veteran. Von Neumann joined Princeton's Institute for Advanced Study (IAS) in 1933, the same year as his mentor, Albert Einstein. Like many of those initially hired by the IAS, von Neumann was a mathematician by training. In 1951, a team of scientists, led by Nicholas Metropolis, constructed a computer called the Mathematical and Numerical Integrator and Calculator (MANIAC). MANIAC's first job was to perform the calculations for the hydrogen bomb. In fact, the RAND Corporation published a book just containing random digits; see Figure 5.9. This is a great read before bedtime!

One of the first ideas to generate random numbers was developed by von Neumann and termed the middle square method. The idea is quite simple:

- Start with a large integer of six digits: this initial value is termed a "seed."
- Square that value.

73735	45963	78134	63873
02965	58303	90708	20025
98859	23851	27965	62394
33666	62570	64775	78428
81666	26440	20422	05720
15838	47174	76866	14330
89793	34378	08730	56522
78155	22466	81978	57323
16381	66207	11698	99314
75002	80827	53867	37797
99982	27601	62686	44711
84543	87442	50033	14021
77757	54043	46176	42391
80871	32792	87989	72248
30500	28220	12444	71840

Figure 5.9 Random digits (RAND Corporation, 1955).

- Retain as output the middle six digits.
- Repeat to get the desired random numbers.

This old method is no longer used, because it generated cycles of random numbers. At some point, the generated value will be equal to the seed, and the cycle starts over. Today, much better methods are available. However, what is common to all methods is the seed. Given a seed, one will always generate the same random number sequence. So, this seed acts like a key to unlock a deterministic series of random numbers. In that sense, scientists prefer to call these series "pseudo-random numbers." What that means is that if you use any computing device (computer/phone) and you do not change the hardware or software configuration, in a million years, with a given seed, you will still get the same series of random values.

Now we are ready to return to our original question: How can a computer program mimic the physical process of picking a random ball out of a bag? To do so we need to return to the empirical cumulative distribution function; see Figure 5.10. We let the pseudo random number generate a single value (imagine the drum with an infinite number of balls containing all real numbers between 0 and 1). Then we put this value, 0.61 in Figure 5.10, on the y-axis, and find the corresponding x-value. Two things happen here:

- Because we go in steps, we can only find one of the sample values.
- Because the y-value is random, you will have a 1/6 chance of hitting one of the steps.

We can extend this idea to very large populations, simply going in small steps. We can go even further and create infinitesimally small steps. That means we can use the same idea to create samples from a model.

Let's emphasize again why this is useful. Often in data science, we like to use theoretical models to study the real world. For example, we can hypothesize that some variable in the real world has the perfect exponential distribution (this is rarely if ever true!). Once we assume this, we can generate sample datasets from it, for example sample datasets with a different sample size. In this way, we can understand the effect of the sample size. Do I have enough samples to make a confident estimate? Or do I need more? One of these ideas is the bootstrap technique.

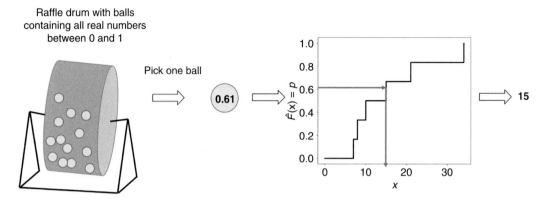

Figure 5.10 Monte Carlo Simulation.

Introduction to the Bootstrap Method

Stanford statistician Bradley Efron transformed statistical science into a computational science by inventing the bootstrap. Most of you grew up with computation right in front of you (your smartphone). This phenomenon is quite new. In the nineteenth and early twentieth century much of science was done without computers. For that reason, statistical models like the exponential distribution were used because of their nice analytical properties. Computing is what allowed us to move away from this and the bootstrap is one such very early method. The name "bootstrap" originates from the tales of Baron von Munchausen, who claimed to have lifted himself up from the swamp by pulling on his own hair.

The problem with parametric models is that they are imperfect reflections of reality, real data. Real data need not follow a specified distribution model; hence, if we assume this, we may be making significant errors. Bootstrapping uses the actual data; no definition of a model is needed! These methods are therefore also termed: non-parametric.

To understand why this is so powerful and how it works, consider a simple problem: estimating the mean of a population. Here we consider a population of diamonds (a very large sample N) that has a certain (finite) mean or arithmetic average. Unfortunately, we do not have access to the full population; instead, we only have a limited sample dataset of size $n << N$. From this limited sample, we can calculate the mean:

$$\hat{\mu} = \bar{x} = \frac{1}{n}\sum_{i=1}^{n} x_i \tag{5.34}$$

One can ask the following questions, all of which are very closely related:

- How "good" is this estimate?
- How close is it to the true value?
- How confident are we in this estimate?
- How can we construct intervals within which the true mean could be located?

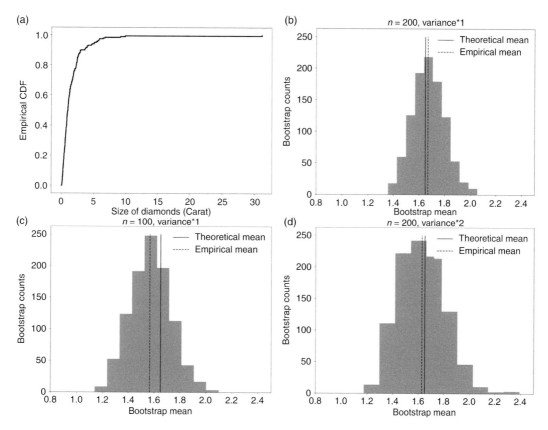

Figure 5.11 (a) Empirical CDF of a sampled dataset ($n = 200$). (b) and (c) Histograms of the bootstrapped means for $n = 200$ (b) and $n = 100$ (c) sampled datasets with the number of bootstraps $b = 1000$. (d) Histogram of the the bootstrapped means on a new $n = 200$ sampled dataset with $b = 1000$. This dataset has twice the variance of the $n = 200$ sampled dataset in (b).

The idea of bootstrap is to perform Monte Carlo simulations using the actual data. We do exactly what we did in the previous section and first create an empirical CDF of the data; see Figure 5.11a. Then we create new datasets as follows:

- We sample n times from the empirical CDF.
- We put those n samples together in a new dataset that we call $\mathbf{x}_b$.
- We repeat this a large number of times (possibly millions), namely b times.

For each of these new datasets we can calculate the mean again. Figure 5.11b shows the result of bootstrap: a histogram of these new estimates of the mean.

Notice how the estimates are now spread out around the original estimate. Let's repeat the same exercise but using only half the original samples n and replot the histogram (Figure 5.11c). We notice that the spread is larger, simply because we are less confident. Let's now return to the above questions. To investigate these, the actual value is also shown. The "goodness" of the estimate depends in fact on two concepts:

- How many samples do we have?
- How much variation is there in the original population?

To check on the last condition, we consider a second diamond dataset whose variance is twice as large. We repeat the same exercise of bootstrap and show the results in a histogram (Figure 5.11d). We notice indeed that when the population has more variation, the confidence in the estimation of the mean decreases.

Bootstrap Confidence Intervals

Sometimes we'd like to summarize the histogram of bootstrap estimates. The best way to proceed is to report some quantiles of the data displayed in Figure 5.11, for example, the 5% and 95% quantiles. These quantiles mark the end-points of what are termed confidence intervals. The interval between the 5% and 95% therefore is a 90% confidence interval. This idea of bootstrapping can be applied to any estimate, not just the mean, hence the generality of the idea.

We obtain the bootstrap confidence intervals of the mean and the variance of the two diamond datasets in Chapter 1 (Figure 1.4). Dataset 1 has 1000 samples and dataset 2 has 500 samples. Figure 5.12 shows that dataset 1 has smaller confidence intervals for both

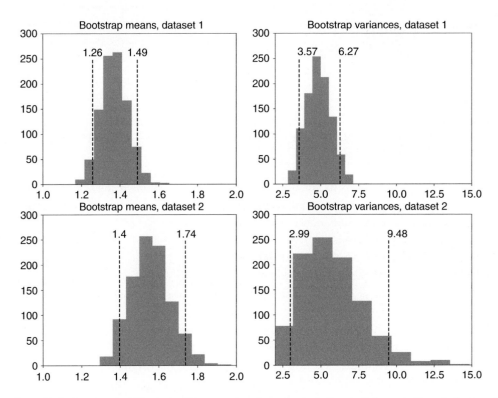

Figure 5.12 Bootstrap estimations of mean and variance for two diamond datasets. Two dashed vertical lines are the 5% and 95% quantiles of bootstrap estimations. 90% bootstrap confidence intervals are between those two quantiles.

bootstrap means and variances. We are more confident on the mean and variance estimation of dataset 1.

Hypothesis Testing Using the Bootstrap Method

Testing a hypothesis is key to many areas of science and engineering. Hypothesis testing starts often by asking important questions: Was there life on Mars? Was area X enriched in minerals? Was agriculture the source of pollution? Hypothesis testing provides a rigorous statistical protocol that may help in addressing these questions. As the term suggests, hypothesis testing consists of two components: stating the hypothesis, then running a test. It is important to do it in that order. Running tests and then trying to confirm some conclusion may not be as objective compared to first clearly stating what is being tested. For that reason, hypothesis tests are also termed hypothetico-deductive ways of reason. That's a complicated way of saying that the way we reach conclusions is to falsify or reject an alternative conclusion. Hypothesis testing is therefore a falsification type test: it aims to disprove the opposite, rather than confirming a statement.

How does hypothesis testing work? Here is a broad protocol:

- State what you want to study: "I want to study the origin of diamonds."
- Relate the scientific study to a statistical study: "Diamonds from kimberlites are distributed as a Pareto distribution, while diamonds of river deposits are lognormal," as observed from previous data.
- Turn the question into a falsifiable hypothesis. "The extreme value index ξ equals 0" (this will reject the Pareto distribution as a model, hence the kimberlite as a source). It is important to be precise in stating a falsifiable hypothesis. "ξ may be 0" or "I think ξ is 0" is not falsifiable! "ξ equals 0" is an example of something falsifiable, because if you know for sure $\xi > 0$, then this hypothesis is falsified.
- Determine how well you know the parameter from limited data. In other words: quantify the uncertainty of that parameter.
- Using the uncertainty quantification of that parameter, reject (or not) the hypothesis.

Let's perform hypothesis testing on the two diamond datasets in Chapter 1. Our falsifiable hypothesis is "The extreme value index ξ equals 0." We use the bootstrap to quantify the uncertainty of the estimate of the extreme value index ξ. ξ is estimated using the linear fit of the mean excess quantile plot (Section 1.5.2). For each round of bootstrap, we bootstrap k values larger than a given threshold u. These k samples might contain repeated values because we performed bootstrap sampling with replacement. Figure 5.13 shows two examples (#1, #2) of bootstraps for each dataset. The estimation of ξ (the fitted slope) is also different for each bootstrap sample. We then perform the bootstrap many times ($b = 10,000$), from which we can calculate a confidence interval for the estimated ξ.

For dataset 1, the 90% confidence interval is: [−0.58, −0.15]. For dataset 2, the 90% confidence interval is: [0.27, 0.37]. Therefore, we reject the hypothesis "The extreme value index ξ equals 0" for both datasets.

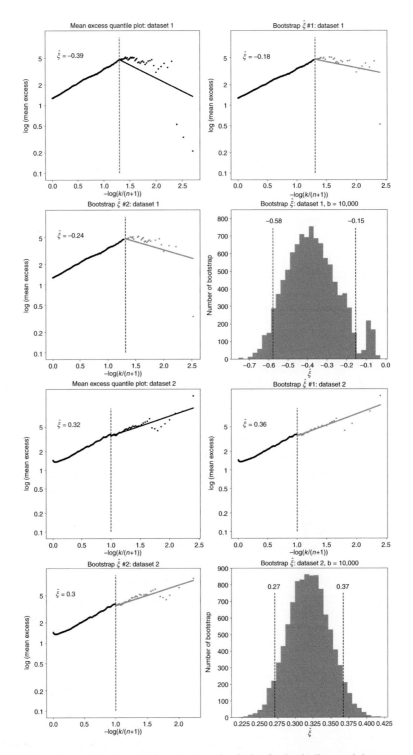

Figure 5.13 Bootstrap estimation of the extreme value index for both diamond datasets.

5.3   Concept Review for Chapter 2: Multivariate Analysis

We briefly mentioned some basic concepts of multivariate analysis in Chapter 2, including data standardization, covariance, rotations, eigenvalue decomposition, and distances versus similarity. Here we review some important basics arounds these concepts in an intuitive fashion.

5.3.1   Standardizing Multivariate Data

Providing summary statistics on a dataset is the first step to provide insights into the data. When dealing with only one variable, we commonly report the mean and standard deviation of the dataset and, as discussed in Chapter 1, possibly some quantiles if we have a skewed distribution. When we have two variables at a time, we can report the mean and standard deviation of each variable, as well as the correlation coefficient. When dealing with more than three variables, things get more complicated, so a first step can be to show a matrix of scatter plots, and a corresponding matrix of correlation coefficients; see Figure 5.14. This matrix has to be symmetric because the correlation coefficients are symmetric as well.

Here we will go a bit deeper into the equation of the correlation coefficient r:

$$r = \frac{\sum_{i=1}^{n}(x_i - \bar{x})(y_i - \bar{y})}{\sqrt{\sum_{i=1}^{n}(x_i - \bar{x})^2 \sum_{i=1}^{n}(y_i - \bar{y})^2}} \tag{5.35}$$

In many applications we want to compare datasets but not in terms of mean and standard deviation. Instead, we may wish to transform two datasets or two variables into a new dataset

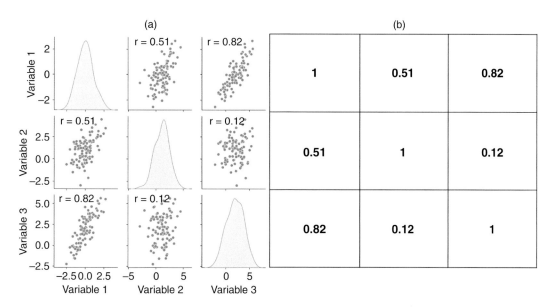

Figure 5.14 Correlation between three variables: (a) pairwise scatter plot off the diagonal and the kernel density of each variable on the diagonal; (b) correlation-coefficient matrix.

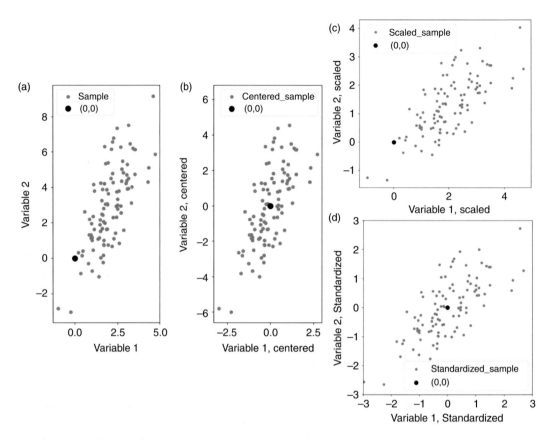

Figure 5.15 Standardizing bivariate data: (a) original bivariate data; (b) centered bivariate data; (c) scaled bivariate data; (d) standardized bivariate data.

such that they all have a mean equal to 0 and a standard deviation equal to 1. Take the example in Figure 5.15a. The black dot is the value (0,0). We would like to move the scatter of points such that the black dot is in the middle of the cloud. This operation is termed "centering" the data. To do this, we calculate the mean of each variable, then subtract the mean from our data.

For a bivariate dataset $\{(x_1, y_1), (x_2, y_2), \ldots, (x_n, y_n)\}$, we calculate the mean of the x values:

$$\bar{x} = \frac{1}{n} \sum_{i=1}^{n} x_i \tag{5.36}$$

and the mean of the y values:

$$\bar{y} = \frac{1}{n} \sum_{i=1}^{n} y_i \tag{5.37}$$

and create a centered dataset:

$$(x_1 - \bar{x}, y_1 - \bar{y}), (x_2 - \bar{x}, y_2 - \bar{y}), \dots, (x_n - \bar{x}, y_n - \bar{y}) \qquad (5.38)$$

We notice that in this new plot, Figure 5.15b, the centered variables are not on the same scale, x goes from -2 to 2, and y goes from -7 to 7. Indeed, in many cases we plot variables with different units (e.g., kg vs meter). One way to remove the unit is to scale the data with the standard deviation. We calculate the standard deviation as:

$$s_x = \frac{1}{n}\sqrt{\sum_{i=1}^{n}(x_i - \bar{x})^2}; s_y = \frac{1}{n}\sqrt{\sum_{i=1}^{n}(y_i - \bar{y})^2} \qquad (5.39)$$

where s_x and s_y are the standard deviations for x and y respectively. Note that in most textbooks you will see a division by $(n-1)$. This is done to make the estimator unbiased; however, dividing by n or $n-1$ will not differ that much and the bias is very small, relative to other uncertainties and errors we make.

When dividing by the standard deviation we get:

$$\left(\frac{x_1}{s_x}, \frac{y_1}{s_y}\right), \left(\frac{x_2}{s_x}, \frac{y_2}{s_y}\right), \dots, \left(\frac{x_n}{s_x}, \frac{y_n}{s_y}\right) \qquad (5.40)$$

The result of this is shown in Figure 5.15c. The next obvious thing to is to center and scale jointly. This operation in statistics is termed "standardization," the new standard is zero mean and unit standard deviation; see Figure 5.15d. Now we also get a bit more insight into the correlation coefficient formula. We notice that in this calculation we have standardized the x and y values. This is the reason that r is between -1 and 1, which makes it easy to interpret. A nice property of this standardization is that it is easy to undo it:

$$b = \frac{a - m}{s} \rightarrow a = s \times b + m \qquad (5.41)$$

where a is the original data value, b is the value after standardization, and m and s are the empirical mean and standard deviation. This means that we can perform a statistical analysis on the standardized data and, when we are done, simply undo the standardization, that is, back-transform.

However, the standardization of compositions cannot involve simply the mean and standardization. That would violate the stated subcompositional coherence requirement, so we need to look for different ways of doing this (Section 2.4.4).

5.3.2 Covariance and Semi-Variograms

In Section 5.2.2 we discussed the notion of theoretical expectation and variance, as quantities characterizing an entire population with a given distribution model, as opposed to the empirical version calculated from limited sample data. Similarly, we can define the population correlation in terms of the expectation E. When defining the population correlation, we deal with two random variables X and Y, each with their own mean m and variance σ^2:

$$m_X = E[X]; m_Y = E[Y]; \sigma_X = \sqrt{var(X)}; \sigma_Y = var(Y) \qquad (5.42)$$

The population or theoretical correlation is defined as:

$$\rho_{XY} = \frac{E[(X - m_X)(X - m_Y)]}{\sigma_X \sigma_Y} \qquad (5.43)$$

Because of the division by σ_X and σ_Y, the correlation is unitless. The unitless property is very useful because we can compare degrees of correlation between say X and Y with X and Z. However, in some applications we do care about the variance, for example, if we want to study contributions of variance to the total variance, or the spatial variation of a phenomenon (such as in Chapter 4). In all cases where the actual variance matters in magnitude, we should use the notion of covariance, which is an unstandardized correlation:

$$cov_{XY} = E[(X - m_X)(X - m_Y)] \qquad (5.44)$$

A final measure of correlation is the semi-variogram defined as:

$$\gamma_{XY} = \frac{1}{2} E[(X - Y)^2] \qquad (5.45)$$

The covariance and correlation coefficient measure a degree a similarity, while the semi-variogram measures dissimilarity (the square of the difference). Notice how the definition no longer involves the means m_X and m_Y. This idea is exploited in Chapter 4 to define measures of spatial correlation.

5.3.3 Properties of Three-Dimensional Rotation

The ellipse and the ellipsoid are used in Chapter 2 as ideal shapes for bivariate and trivariate datasets, where the variables in these datasets have a linear relationship and each variable a Gaussian distribution. While this combination of linearity and Gaussianity usually does not occur in most geoscientific applications, it is still useful to develop methodologies starting from those assumptions. Consider, for example, Figure 2.14, a case where the correlation between three variables does look like an ellipsoid. We will now study two questions:

- How do we represent the angles of an ellipsoid?
- How can we use that representation to change the ellipsoid into a sphere?

The usefulness of the latter will be studied in the next section using eigenvalue decomposition, which is the basis of principal component analysis. Turning an ellipsoid into a sphere essentially entails removing the correlation between variables.

Let's first look into the angle representations of an ellipsoid. We take a three-dimensional ellipsoid example in Figure 5.16a. This ellipsoid can rotate independently along three different angles. The first is a rotating angle ψ around the x-axis. We define coordinates of x as $[x, y, z]$, before rotating, and $[x(\psi), y(\psi), z(\psi)]$, is for after rotation. We can calculate the new coordinates as follows:

$$x(\psi) = x$$
$$y(\psi) = y * \cos \psi - z * \sin \psi$$
$$z(\psi) = y * \sin \psi + z * \cos \psi$$

or

$$\begin{bmatrix} x(\psi) \\ z(\psi) \\ y(\psi) \end{bmatrix} = \begin{bmatrix} 1 & 0 & 0 \\ 0 & \cos \psi & -\sin \psi \\ 0 & \sin \psi & \cos \psi \end{bmatrix} \begin{bmatrix} x \\ y \\ z \end{bmatrix} = \mathbf{x} R_x(\psi) \tag{5.46}$$

Rotation constitutes a linear transformation with a matrix operator

$$R_x(\psi) = \begin{bmatrix} 1 & 0 & 0 \\ 0 & \cos \psi & -\sin \psi \\ 0 & \sin \psi & \cos \psi \end{bmatrix} \tag{5.47}$$

$R_x(\psi)$ is called the rotation matrix.

Similarly, the second angle θ is for rotation around the y-axis, with its rotation matrix as

$$R_y(\theta) = \begin{bmatrix} \cos \theta & 0 & \sin \theta \\ 0 & 1 & 0 \\ -\sin \theta & 0 & \cos \theta \end{bmatrix} \tag{5.48}$$

The third rotating angle φ is around the z-axis, and its rotation matrix is

$$R_z(\varphi) = \begin{bmatrix} \cos \varphi & -\sin \varphi & 0 \\ \sin \varphi & \cos \varphi & 0 \\ 0 & 0 & 1 \end{bmatrix} \tag{5.49}$$

ψ, θ, φ are termed the Euler angles. Any rotation can be decomposed into a sequence of rotating movements with the three Euler angles. We therefore have a generalized form of rotation matrix R, which is the multiplication product of $R_x(\psi)$, $R_y(\theta)$, and $R_z(\varphi)$, depending on the order of rotating axis. For example, if we first rotate an object around the x-axis, then the y-axis, and finally the z-axis, the rotation matrix will be

$$R = R_z(\varphi) R_y(\theta) R_x(\psi) \tag{5.50}.$$

For the ellipsoid in Figure 5.16a, if we rotate this ellipsoid first about the x-axis by 45 degrees ($\psi = 45$), then around the y-axis by 30 degrees ($\theta = 30$), and finally around the z-axis by 45

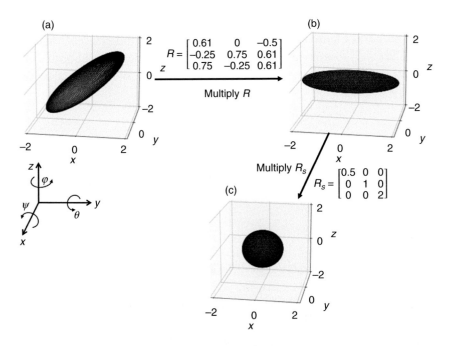

Figure 5.16 Transformation of a three-dimensional ellipsoid to a sphere. (a) The original ellipsoid. Its major, median, and minor axes are (2, 1, 0.5). (b) The ellipsoid is rotated to make the x-axis its major axis, the y-axis its median axis, and the z-axis the minor axis. (c) The rotated ellipsoid is scaled to a sphere with radius = 1.

degrees ($\varphi = 45$), we will get a new ellipsoid in Figure 5.16b. We can calculate the rotation matrix using Eq. (5.50) above, obtaining

$$R = \begin{bmatrix} 0.61 & 0 & -0.5 \\ -0.25 & 0.75 & 0.61 \\ 0.75 & -0.25 & 0.61 \end{bmatrix}$$

What do you observe after the rotations? The x-, y-, and z-axes represent perfectly the major, median, and minor axes of the ellipsoid.

Once the ellipsoid is rotated in Figure 5.16b, we can turn it into a sphere with radius equal to "1" (Figure 5.16c). This is simply done by reducing the major axis by half, increasing the minor axis by two, while keeping median axis unchanged. Mathematically, this means multiplying the ellipsoid's coordinates with a scale factor in each axis (thus a scaling matrix R_s in Figure 5.16).

We learn that any ellipsoid can be represented by a sphere after a series of rotations and scaling. These rotations and scaling are all linear operations as shown in Figure 5.16, known as the "eigenvalue decomposition," discussed in the next section.

5.3.4   Eigenvalue Decomposition of the Covariance Matrix

Eigenvalue decomposition is an essential operation in data science and signal or image processing techniques in general. Here we focus on its use in correlation analysis. Consider the blue jay

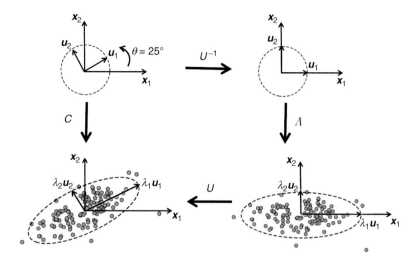

Figure 5.17 Illustration of eigenvalue decomposition using the blue jay dataset from Chapter 2. The x_1 and x_2 scatter plots are the head length and skull size centered at the mean, respectively.

dataset of Chapter 2, shown in Figure 5.17. We calculate the variance–covariance matrix C of the variables x_1 and x_2 using Eq. (2.32)

$$C = \begin{bmatrix} var(x_1) & cov(x_1, x_2) \\ cov(x_2, x_1) & var(x_2) \end{bmatrix} = \begin{bmatrix} 2.46 & 0.96 \\ 0.96 & 0.83 \end{bmatrix} \tag{5.51}$$

The relationship between these two variables can be represented by an ellipse. In the previous section we saw how an ellipse (correlation) can be transformed into a circle (no correlation). To do this, we need only two operations: rotation and squeezing. Let's start the other way around and transform a circle into an ellipse.

   Figure 5.17 shows the transformation. We represent the circle in the original coordinates of blue jay data (x_1 and x_2). u_1 and u_2 are unit vectors indicating the major and minor axes of the data ellipse. The angle between the major axis u_1 and x_1 is 25 degrees ($\theta = 25°$). If we first inversely rotate this circle (with -25 degrees), then stretch it by λ_1 in the u_1 direction and by λ_2 in the u_2 direction, and finally back-rotate by 25 degrees, we obtain the ellipse that represented the data covariance matrix. We can formulate this operation using the two-dimensional rotation matrix $R(\theta)$:

$$R(\theta) = \begin{bmatrix} \cos\theta & -\sin\theta \\ \sin\theta & \cos\theta \end{bmatrix} = \begin{bmatrix} 0.91 & -0.42 \\ 0.42 & 0.91 \end{bmatrix} \tag{5.52}$$

$$C = R(\theta) \begin{bmatrix} \lambda_1 & 0 \\ 0 & \lambda_2 \end{bmatrix} R(\theta)^{-1} \tag{5.53}$$

In linear algebra, λ_1 and λ_2 are termed eigenvalues. The coordinate vectors u_1 and u_2 are termed eigenvectors; u_1 and u_2 are the ellipse major and minor directions, meaning the directions of the data's largest and smallest variances; and λ_1 and λ_2 are the corresponding largest and smallest variances.

We can define the eigenvector matrix $[\boldsymbol{u}_1, \boldsymbol{u}_2]$ as U. The eigenvectors can be calculated using a rotation matrix because they are rotated from the original data coordinate vectors $\begin{bmatrix} 1 & 0 \\ 0 & 1 \end{bmatrix}$:

$$U = [\boldsymbol{u}_1, \boldsymbol{u}_2] = R(\theta) \begin{bmatrix} 1 & 0 \\ 0 & 1 \end{bmatrix} = \begin{bmatrix} 0.91 & -0.42 \\ 0.42 & 0.91 \end{bmatrix} \tag{5.54}$$

The eigenvector matrix U is the same as the rotation matrix! Thus, we can we write Eq. (5.53) as

$$C = U \Lambda U^{-1} \tag{5.55}$$

where $\Lambda = \mathrm{diag}(\lambda_1, \lambda_2)$. Equation (5.55) is called the eigenvalue decomposition of the covariance matrix. This is a two-dimensional example but the same rule applies to higher dimensions. In Chapter 2, we show how this eigenvalue decomposition of the variance–covariance matrix is at the foundation of principal component analysis. The rotation matrix from the eigenvectors of the variance–covariance matrix determines the projection of the PCA.

5.3.5 Distance and Similarity

In multivariate analysis, associations can be viewed in two different ways: the correlation between variables under study and the difference between multivariate samples. Indeed, if the random variables in a multivariate analysis exhibit a high correlation, the samples will be very similar. A distance is a mathematical function that allows us to define similarities (or differences) between any random objects (vectors, shapes, surfaces). However, not all functions are distance functions; we need some rules. First, let's start with some notations.

Consider a set of random variables, and arrange them in a vector. The latter is called a random vector, which we write as $\mathbf{X}$. Like random variables, we use a capital to denote that these are unknown. The dimension of $\mathbf{X}$ is represented by N, the length of the vector. Any particular outcome, such as a sample value, is written as x. Next, we need a notation for multiple samples; here, we use the following: $X = (\mathbf{x}^{(1)}, \ldots, \mathbf{x}^{(\ell)}, \ldots, \mathbf{x}^{(L)})$, $\ell = 1, \ldots, L$. X here is a matrix, it contains L samples of vectors with dimension N, so it is an $N \times L$ matrix.

The following axioms defined a distance:

$$K = \begin{aligned} d\left(\mathbf{x}^{(\ell)}, \mathbf{x}^{(\ell')}\right) &\geq 0 && \text{non-negativity} \\ d\left(\mathbf{x}^{(\ell)}, \mathbf{x}^{(\ell')}\right) &= 0 \Leftrightarrow \mathbf{x}^{(\ell)} = \mathbf{x}^{(\ell')} && \text{identity} \\ d\left(\mathbf{x}^{(\ell)}, \mathbf{x}^{(\ell')}\right) &= d\left(\mathbf{x}^{(\ell')}, \mathbf{x}^{(\ell)}\right) && \text{symmetry} \\ d\left(\mathbf{x}^{(\ell)}, \mathbf{x}^{(\ell'')}\right) &\leq d\left(\mathbf{x}^{(\ell)}, \mathbf{x}^{(\ell')}\right) + d\left(\mathbf{x}^{(\ell')}, \mathbf{x}^{(\ell'')}\right) && \text{triangular inequality} \end{aligned} \tag{5.56}$$

where d is the distance function and $\mathbf{x}^{(\ell)}, \mathbf{x}^{(\ell')}, \mathbf{x}^{(\ell'')}$ are different sampled vectors. Well-known metric spaces are real numbers with absolute difference or any Euclidean space with a Euclidean distance defined as:

$$d\left(\mathbf{x}^{(\ell)}, \mathbf{x}^{(\ell')}\right) = \sqrt{\left(\mathbf{x}^{(\ell)} - \mathbf{x}^{(\ell')}\right)^T \left(\mathbf{x}^{(\ell)} - \mathbf{x}^{(\ell')}\right)} \tag{5.57}$$

One advantage of using distances is that we can use different distance definitions. We are not restricted to the Euclidean distance. A wide class of differences are generated based on p-norms:

$$d\left(\mathbf{x}^{(\ell)}, \mathbf{x}^{(\ell')}\right) = \|\mathbf{x}^{(\ell)} - \mathbf{x}^{(\ell')}\|_p = \left(\sum_{n=1}^{N} |\mathbf{x}^{(\ell)} - \mathbf{x}^{(\ell')}|^p\right)^{1/p} \tag{5.58}$$

An assumption here is that the components of $\mathbf{x}$ are on the same scale. The Manhattan norm ($p = 1$) gives rise to the Manhattan distance (also termed ℓ_1-distance), where the distance between any two vectors is the sum of the differences between corresponding components. The maximum norm (ℓ_∞-distance) gives rise to the Chebyshev distance or chessboard distance where $p = \infty$.

The norms above are appropriate when dealing with continuous variables; however, they become problematic for categorical variables S, which may not have ordinality. Consider the following examples of geological sequences:

$$\mathbf{s}^{(1)}: D\ F\ F\ E\ D\ E \qquad\qquad D = delta, \quad F = fluvial, \quad E = estuarine$$

$$\mathbf{s}^{(2)}: F\ D\ F\ E\ F\ E$$

What is the measure of their difference? In the absence of order, there should be no difference between D and E or D and F. To alleviate this, we need to create indicator variables I:

$$I_D(s) = \begin{cases} 1 & if \ \ s = D \\ 0 & else \end{cases} \quad I_E(s) = \begin{cases} 1 & if \ \ s = E \\ 0 & else \end{cases} \quad I_F(s) = \begin{cases} 1 & if \ \ s = F \\ 0 & else \end{cases} \tag{5.59}$$

where S is a discrete random variable that takes one of three categories $\{D, E, F\}$, s is a sample of the random variable S, $\mathbf{S}$ is a random vector with a set of random variables S, and $\mathbf{s}^{(1)}$ and $\mathbf{s}^{(2)}$ are samples of the random vector $\mathbf{S}$.

With the three new indicators in Eq. (5.59), we rewrite geological sequences $\mathbf{s}^{(1)}$ as $I_D\left(\mathbf{s}^{(1)}\right) = [1\ 0\ 0\ 0\ 1\ 0]$, $I_E\left(\mathbf{s}^{(1)}\right) = [0\ 0\ 0\ 1\ 0\ 1]$, $I_F\left(\mathbf{s}^{(1)}\right) = [0\ 1\ 1\ 0\ 0\ 0]$, $\mathbf{s}^{(2)}$ as $I_D\left(\mathbf{s}^{(2)}\right) = [0\ 1\ 0\ 0\ 0\ 0]$, $I_E\left(\mathbf{s}^{(2)}\right) = [0\ 0\ 0\ 1\ 0\ 1]$, $I_F\left(\mathbf{s}^{(2)}\right) = [1\ 0\ 1\ 0\ 1\ 0]$. We represent $\mathbf{s}^{(1)}$ and $\mathbf{s}^{(2)}$ using sequences of 0 or 1.

Using the Euclidean distance in Eq. (5.57), an appropriate distance is

$$d\left(\mathbf{s}^{(1)}, \mathbf{s}^{(2)}\right) = \frac{1}{3} \sum_{s \in \{D,E,F\}} \sqrt{\left(I_s\left(\mathbf{s}^{(1)}\right) - I_s\left(\mathbf{s}^{(2)}\right)\right)^T \left(I_s\left(\mathbf{s}^{(1)}\right) - I_s\left(\mathbf{s}^{(2)}\right)\right)} \tag{5.60}$$

Finally, we can also define distances that specialize in defining similarities between two histograms, or two PDFs.

Consider, to that end, two discrete probability distributions represented by the discrete probabilities $p_k, q_k, k = 1, \ldots, K$, with K different categories. A well-known distance is the chi-square distance:

$$d_{\chi^2}(\mathbf{p}, \mathbf{q}) = \frac{1}{2} \sum_{k=1}^{K} \frac{(p_k - q_k)^2}{(p_k + q_k)} \tag{5.61}$$

This distance may underweight small differences due to the square, but it is symmetric. Another measure of difference related to information theory is the Kullback–Liebler (KL) divergence:

$$dif\, f_{\mathrm{KL}}(\mathbf{p}, \mathbf{q}) = \sum_{k=1}^{K} p_k \log \frac{p_k}{q_k} \tag{5.62}$$

which is the expected value of the logarithmic differences. This measure is not symmetric, hence not a distance. The symmetric form of the KL divergence is the Jensen–Shannon (JS) divergence:

$$d_{\mathrm{JS}}(\mathbf{p}, \mathbf{q}) = \frac{1}{2} dif\, f_{\mathrm{KL}}(\mathbf{p}, \mathbf{q}) + \frac{1}{2} dif\, f_{\mathrm{KL}}(\mathbf{p}, \mathbf{q}) \tag{5.63}$$

Note that the continuous form of the KL divergence is

$$dif\, f_{\mathrm{KL}}\big(p(x), q(x)\big) = \int p(x) \log \frac{p(x)}{q(x)} dx \tag{5.64}$$

with $p(x)$ and $q(x)$ densities. What is interesting about the KL and JS divergences is that they use a log ratio between two probabilities or two densities. In Chapter 2, we discuss deeper why this is. A probability sums to 1, which we call a composition. A key element of compositions is that calculations involve log ratios.

5.4   Concept Review for Chapter 3: Spatial Data Aggregation

We used probability throughout the entire Chapter 3. What are the general rules of probability that we should always follow? Thomas Bayes established the "kingdom" of Bayes' rule. It governs broad aspects of data science, particularly spatial data aggregation. What really motivated him? In this section, we'll study these two questions.

5.4.1   The Rules of Probability

Mathematics operates through logic and rules around logic. Otherwise, we end up with everyone using their own intuition. Science progresses that way. Rules are made, experiments done, theories postulated. Eventually, theories and rules will change because of new insights, new data that arise. The notion of probability has been around for a while and goes back to a gambler's dispute between the French mathematicians Blaise Pascal and Pierre de Fermat in 1654. The dispute concerned whether an even bet should be placed on throwing at least one double six with two dice in 24 attempts. Since then, many have contributed to developing the theory (including Von Mises, Chebyshev, and Markov), but it was Andrey Kolmogorov who, in 1933, outlined an axiomatic approach to probability. Axioms (from the Greek axíoma) are statements taken to be true, often because they are self-evident. These axioms are then used, in conjunction with more definitions, to produce theories, which need to be proven. What are these self-evident facts we can produce about probabilities?

- *Axiom of positivity*: The probability of an event is a positive value.
- *Axiom of unitarity*: The sum of probability of all possible events is one. The "one" is a choice that makes for ease of calculation; we could have chosen "two"! This axiom is like saying "the universe exists." Now, we are not going to be that bold; instead we will first define our own "universe" of possibilities, such as, "it rains today" and "it does not rain today" is the universe of option of "raining today." The most important part of this axiom is therefore to define the universe within which we work.
- *Axiom of mutual exclusion*: this one is more tricky but very important. First, we need to agree on what are mutually exclusive events: for example E_1 "it rains tomorrow" and E_2 "it rains today" do not "overlap," while "it rains today" and "it rains this afternoon" do overlap. Mutual exclusion does not say anything about how dependent events are, just that there is no overlap. It also means that you cannot use "it may rain tomorrow," which is not a verifiable statement. The consequence of mutual exclusion is that you can sum probabilities when using the "or" statement, namely the probability that "it rains today" *or* "it rains tomorrow" is simply the probability of "it rains today" *plus* "it rains tomorrow."

Following these rules, definitions were made, more specifically the definition of conditional probability which we cover extensively in Chapter 3.

5.4.2   What Motivated Thomas Bayes?

Thomas Bayes was a statistician, philosopher, and reverend. He presented a solution to the problem of inverse probability in "An Essay towards solving a Problem in the Doctrine of Chances." This essay was edited by Richard Price for the Royal Society of London, a year after Bayes' death. Bayes' theorem remained in the background until it was reprinted in 1958. We cover one aspect that Bayes was puzzled about; again, like Monte Hall, it was a game. This new game, billiards, helps to introduce a few fundamental concepts in probability as well.

Prior to that of Bayes, most works on chance were focused on calculating probability using frequencies, such as the number of replications needed to calculate a desired level of probability (how many flips of the coin are needed to assure 50/50 chance?). Bayes considered the problem of inverse probability: "given the number of times an unknown event has happened and failed required that the probability of it happening in a single chance lies between any two degrees of probability that can be named." That's a very long way of saying: if you observe the odds in a game (win/lose) and, say, won five out seven times, do you know what the probability associated with this game is?

What is Bayes' game of interest? Imagine a room with a billiard table; there is another person and one black ball and one white ball. You are asked to go outside the room. The person in the room then tosses the white ball onto the table. This ball stops at a location. The ball now divides the table into two halves, a half to one side, a half to the other side of the ball. Say the length of the table is 1 and the left half has length p, therefore the right half has length $1 - p$. You do not know p because you are in the other room, but otherwise you know what is happening. Now this person takes the black ball and tosses it over the table. This black ball will then land either in the left half, with probability p, or the right half, with probability $1 - p$. When it lands on the left half, you win; otherwise, you lose. Now imagine we have tossed the black ball eight times, you won five times, your opponent won three times, the game is to get to six first. Simple question: What is the probability you will win? What are your odds? 2:1, 5:1, 10:1?

We are reasoning about two probabilities: the probability that you win the game, which means that you need to win only once more; your opponent will win if they win all three times. The second probability concerns the probability of winning when playing only one time. In fact, the probability you lose (the other wins) equals $(1 - p)^3$, hence the probability that you win is $1 - (1 - p)^3$. Therefore, the odds of you winning are

$$\frac{1 - (1 - p)^3}{(1 - p)^3} \tag{5.65}$$

The only remaining question: What is p?

The Binomial Distribution

Various ways of reasoning existed about problems like these. A first idea is to acknowledge that we don't know p, so identifying $p = 0.5$ was in fact considered as a logical choice: it is the least knowledge. If we plug that into our odds, we find

odds of you winning $= 7 : 1$

However, there is an error in logic concerning this solution: we don't know p, therefore we take it to be exactly equal to 0.5 (not 0.51 or 0.49!); plus we already have information, namely we played the game a few times, and it seems this information should not be ignored.

What do we know about p? To understand this, we need a new concept and that is what happens when we do repeated experiments with a given probability, which is what tossing the black ball means. Suppose we know the true p, then we can predict what will happen: out of the eight times, we could have any amount, namely 0, 1, . . . 8 black balls on one side, and we can calculate the probability associated with each situation. Those situations are as follows:

0 balls left vs 8 balls right; 1 ball left vs 7 balls right; . . . 8 balls left vs 0 balls right

You will agree that the probabilities of any of these possible outcomes depend on p. If $p = 0.5$, then "0 balls left vs 8 balls right" will be unlikely, but not zero! Here is the formula:

$$P(k \text{ balls left}, n - k \text{ balls right}) = \binom{n}{k} p^k (1-p)^{(n-k)} \tag{5.66}$$

Let's unpack this equation. The probability that a player wins for one try is p, so if they win three time happens with probability $p \times p \times p$, it also means that losing five times is equated with $(1-p)^5$. However, the order in which the win/lose happens does not matter. Only the total matters, so we have the term in the beginning that describes this amount. This term is important because there is only one possible order for 0–8, while there are eight orders for 1–7. The formula is also better known as the binomial distribution, the "binomial" refers to the two terms, a term p and a term $1-p$. $\binom{n}{k} = \dfrac{n!}{k!(n-k)!}$ is the total number of combinations for having exactly k successes of the event with the probability p in n trials, which is also called the binomial coefficient.

Maximum Likelihood

If we know p, how likely is it to get a score of 5 : 3 in your favor? Now we know the answer lies in the binomial formula:

Probability of getting a score of 5:3 in a game with p

$$= P(5 : 3|p) - \binom{8}{5} p^5 (1-p)^3 \tag{5.67}$$

There are many values of p that can produce the score 5 : 3. So, if we want to settle on one single p, we need an additional criterion. That criterion in statistics is termed the *maximum likelihood*: find that p for which 5 : 3 is the most likely outcome among all possible choices for p. Mathematically, this means that we need to find the p that maximizes $P(5 : 3|p)$. We can find this maximum analytically, first by taking the logarithm, which does not affect the maximum since it is a monotonic function:

$$\left(5 \log p + 3 \log (1-p) + \log \left(\binom{8}{5} \right) \right) \tag{5.68}$$

To find a maximum of a function we need to take the derivative and set it to zero, or

$$\frac{5}{p} + \frac{3}{p-1} = 0 \Rightarrow p = \frac{5}{8} \tag{5.69}$$

That's indeed a very simple and intuitive solution, we could have thought of that without any of the mathematics! That may be true in this case, but the idea of maximum likelihood as applied here can be extended to all kinds of other situations, such as using a dataset to find parameters of a probability model.

We now have the most likely p, so we can use this to predict our odds of winning using the binomial distribution; we find:

- probability of winning $= 1 - (1 - p)^3 = 0.947$
- probability of losing $= (1 - p)^3 = 0.053$

Therefore, the odds are about 18 : 1.

The Bayesian Solution

So far, we have predicted our odds by making a single guess on p. However, if we simulated this game on a computer, we would find that the actual odds are close to 10 : 1. What went wrong? The most consequential decision we made so far in our analysis was to use a single p. Logically, we know this is not true: we know there is a true p, and it is likely different from $5/8$. Additionally, we know that many values of p exist that can produce a score of 5 : 3. There is also a third component to this: we know a billiard table is flat, so we have a priori knowledge, meaning that before the game starts we can reasonably assume p could be any value between 0 and 1. Mathematically, this means that the probability density function of p a priori is a uniform distribution over the interval $[0, 1]$. Now we can invoke Bayes' rule (recall Eq. (3.10) in Chapter 3):

- prior distribution $f(p) = 1$ on $[0, 1]$
- likelihood distribution: $P(5 : 3|p) = \binom{8}{5}p^5(1 - p)^3$
- total probability rule:

$$P(5 : 3) = \int_0^1 \binom{8}{5}p^5(1 - p)^3 f(p)\mathrm{d}p = \int_0^1 \binom{8}{5}p^5(1 - p)^3 \mathrm{d}p \tag{5.70}$$

- posterior distribution:

$$f(p|5 : 3) = \frac{P(5 : 3|p)f(p)}{P(5 : 3)} = \frac{\binom{8}{5}p^5(1 - p)^3}{\int_0^1 \binom{8}{5}p^5(1 - p)^3 \mathrm{d}p} \tag{5.71}$$

- probability of you losing (meaning the other person wins three times in a row)

$$p_{win} = \int_0^1 (1 - p)^3 f(p|5 : 3)\mathrm{d}p = \frac{\int_0^1 p^5(1 - p)^6 \mathrm{d}p}{\int_0^1 p^5(1 - p)^3 \mathrm{d}p} = \frac{1}{11} \tag{5.72}$$

- probability of you winning: $1 - p_{win}$
- odds: 10:1

The correct solution is not as trivial as we perhaps thought and that's why Bayes' rule is so powerful. But you do notice that we need much more calculation with Bayes' rule compared to maximum likelihood, where the only thing we had to do was to divide 5 by 8. This is a common issue that arises when using the comprehensive Bayesian framework and that is many calculations. Here, they are done analytically, and the problem involves only one parameter: p. You can imagine that, with many more parameters, the computations become much more challenging.

However, the logic behind Bayes' idea to solve the billiard problem is surprisingly simple. We assume a p, we play the game, calculate the likelihood of having a score 5:3 under p, and then repeat this for all p.

REFERENCES

RAND Corporation (1955). *A Million Random Digits with 100,000 Normal Deviates.* https://en .wikipedia.org/wiki/A_Million_Random_Digits_with_100,000_Normal_Deviates

Lemieux, C. (2009). *Monte Carlo and Quasi-Monte Carlo Sampling.* Springer. https://doi.org/10.1007/978-0-387-78165-5

Silverman, B. W. (1986). *Density Estimation for Statistics and Data Analysis.* Chapman and Hall.

Index

Printed in the United States
by Baker & Taylor Publisher Services